BRIDGING THE DIGITAL DIVIDE IN THE US

The US faces a deep digital divide which cuts across both urban and rural lines, but is most marked in rural, low-income, and minority communities. This book presents a multi-level governance framework which explores how local leaders use policy opportunity and community resilience to address inequity in broadband infrastructure. Cases from communities across the US are profiled to show how local and regional initiatives address the digital divide – how they harness local resources, state and federal funding, and how they navigate regulatory restrictions and financial constraints. Special attention is given to rural and Indigenous communities, where the technological, organizational, and funding challenges are greatest. The federal policy landscape for broadband is changing, and this book provides clear insights on how policy can support the people and places left behind. This book is essential for planners and those studying or working in rural development, rural studies, and community development.

Mildred E. Warner is a Professor of City and Regional Planning and of Global Development at Cornell University. She is an expert on state and local government policy and has published widely on infrastructure, economic development, finance, service delivery, and community development.

Natassia A. Bravo received her Ph.D. in City and Regional Planning from Cornell University. Her research focuses on infrastructure policy, especially rural broadband access. She has published popular reports on broadband policy and local initiative. She and Dr. Warner were winners of the Charles Benton

Broadband & Society Prize in 2023. Her research focuses on infrastructure policy, especially rural broadband access, and she has published popular reports on broadband policy and local initiative.

Duxixi (Ada) Shen is an infrastructure planning consultant and former researcher at Cornell whose research focuses on bridging the digital divide in a wide array of communities, including Colorado, Maine, Alaska, and the Choctaw Indian Nation. She received her master's degree in Regional Planning from Cornell.

"Bringing broadband to unserved Americans is not as simple as hanging wire on a pole. While most literature focuses on a single aspect of the digital divide–market dynamics, availability of middle mile, impact measurement – Warner, Bravo, and Shen have brought those pieces together in a single compendium, clarifying our understanding of the problem and deepening our understanding of solutions. This book fills a critical research gap and is an essential read for any practitioner of telecommunications, infrastructure, state or local policy."

Kathryn de Wit, *Project Director, Broadband Access Initiative,*
The Pew Charitable Trusts

"This book offers a fresh perspective on broadband policy, emphasizing the intricate interplay between federal, state, and local efforts to close the digital divide. It is an essential read for anyone interested in advancing digital equity and for scholars exploring the dynamics of multilevel governance in the 21st century."

Hernan Galperin, *Professor and Director of Doctoral Studies,*
USC Annenberg School for Communication,
University of Southern California

"This book explores the next chapter in U.S. Digital Divide research, taking us into how communities can build durable programs to redress digital needs and inequities. *Closing the Digital Divide in the US: Planning Innovative State and Local Approaches* should be required reading for the groups and policymakers who want permanent solutions to delivering the online connectivity everyone requires in the 21st century."

Sharon Strover, *Philip G. Warner Regents Professor, University of*
Texas at Austin and Co-Director, Technology and Information
Policy Institute

"*Bridging the Digital Divide in the US* lays out a comprehensive roadmap on how to achieve universal broadband access and digital equity. This collection addresses a missing piece in the digital divide scholarship by describing how state and local initiatives can effectively address social and digital inequity."

Colin Rhinesmith, *Director, Digital Equity Action Research (DEAR)*
Lab, School of Information Sciences, University of Illinois
Urbana-Champaign

"This is an important book for a time when the federal government is considering dramatic changes to Internet policy. It provides keen insight into the local and regional programs and partnerships that are making a difference for people on the wrong side of the divide. Many thoughtful examples provide inspiration and details for other communities to follow."

Christopher Mitchell, *Director, Community Broadband Networks, Institute for Local Self-Reliance*

"With thoughtful policy analyses and insightful case studies, this is the first edited collection to rightly highlight and champion the crucial role played by local and state initiatives and stakeholders working to make digital equity a reality in the United States."

Dr. Christopher Ali, *Pioneers Chair in Telecommunications, Penn State University*

The Community Development Research and Practice Series

Series Editor:

Mark Brennan, Pennsylvania State University, USA

Editorial Board:

Norman Walzer, Northern Illinois University, USA

Jean Lonie, Tarleton State University, USA

Brad Olsen, University of California, Santa Cruz, USA

Sebastian Galindo, University of Florida, USA

Zach Wood, Seattle University, USA

Tanja Hernandez, Penn State University, USA

Other books in the series:

For more information about this series, please visit: https://www.routledge.com/ Community-Development-Research-and-Practice-Series/book-series/CDRP

BRIDGING THE DIGITAL DIVIDE IN THE US

Planning Innovative State and Local Approaches

Edited by
Mildred E. Warner, Natassia A. Bravo
and Duxixi (Ada) Shen

Routledge
Taylor & Francis Group
NEW YORK AND LONDON

Designed cover image: Shutterstock

First published 2026
by Routledge
605 Third Avenue, New York, NY 10158

and by Routledge
4 Park Square, Milton Park, Abingdon, Oxon, OX14 4RN

Routledge is an imprint of the Taylor & Francis Group, an informa business

Open Access funding provided by the USDA National Institute of Food and Agriculture, Agricultural and Food Research Initiative Competitive Program, Agriculture Economics and Rural Communities, grant #2021-67023-34437, "Local Government, Inequality and Rural Wellbeing."

ISBN: 9781041024262 (hbk)
ISBN: 9781032914602 (pbk)
ISBN: 9781003619208 (ebk)

DOI: 10.4324/9781003619208

Typeset in Times New Roman
by codeMantra

*This book is dedicated to all the local leaders who work
so hard to bring broadband to their communities.*

CONTENTS

FIGURES

TABLES

CONTRIBUTORS

Dr. Johannes M. Bauer (Ph.D., Economics) is the Quello Chair of the James H. and Mary B. Quello Center at Michigan State University. From September 2023 through December 2024, he served as Chief Economist, U.S. Federal Communications Commission (FCC). He has published extensively on digital equity issues.

Dr. Natassia A. Bravo (Ph.D., City and Regional Planning) received her degree from Cornell University. Her research focuses on infrastructure policy, especially rural broadband access. She has published popular reports on broadband policy and local initiative. She and Dr. Warner were winners of the *Charles Benton Broadband & Society Prize* in 2023.

Dr. Roberto Gallardo (Ph.D., Public Policy) is Vice President for Engagement and Associate Professor in the Agricultural Economics Department at Purdue University, where he also serves as Senior Associate and Advisor to the Center for Regional Development. Dr. Gallardo has published widely on issues of regional development and digital transformation.

Dr. John J. Green (Ph.D., Rural Sociology) is Director of the Southern Rural Development Center, one of the nation's four Regional Rural Development Centers, and Professor in the Department of Agricultural Economics at Mississippi State University. His areas of interest include population change, rural community development, digital inclusion, and capacity building.

Yixiao Edward Guo (Master's, City and Regional Planning) focuses on infrastructure and how communities collaborate to address the digital divide. He has worked for Ithaca Area Economic Development to help develop a rural broadband extension project in Tompkins County, NY. He received his master's degree from Cornell University.

Elizabeth H. Redmond (Master's, Regional Science) focuses on the links between critical infrastructure and regional development. Her work has focused on innovative approaches to broadband in rural communities. She received her master's degree from Cornell University.

Dr. Roseanne Scammahorn (Ph.D., Human Sciences) is Associate Director of the Southern Rural Development Center, where she conducts applied research to inform outreach and education curricula for capacity building, including digital skills. She has experience working as an educator in several roles within the Cooperative Extension System.

Duxixi (Ada) Shen (Master's, City and Regional Planning) is an infrastructure planning consultant and former researcher at Cornell, whose research focuses on bridging the digital divide in a wide array of communities, including Colorado, Maine, Alaska, and the Choctaw Indian Nation. She received her master's degree in Regional Planning from Cornell.

Dr. Mildred E. Warner (Ph.D., Development Sociology) is a Professor of City and Regional Planning and of Global Development at Cornell University. She is an expert on state and local government policy and has published widely on infrastructure, economic development, finance, service delivery, and community development.

PREFACE

Over the past few decades, our society has undergone a massive digital transformation – the Internet now appears nearly everywhere. Movies, books, and the process of socialization, in general, have been moved on to or made drastically different by the Internet. Remote work, online learning, e-commerce, and tele-health, all enabled via the Internet, can bring users great flexibility and convenience. But for many, the digital revolution has left them behind – excluded and marginalized. There is a deep digital divide in the US, and for those on the other side, accessing information and services can be almost impossible.

The disruption caused by the COVID-19 pandemic furthered our dependency on the Internet as school, work, and even governmental affairs shifted online. But it also exposed the digital divide for everyone to see. Many children and adults, especially in rural, low-income, and minority areas, struggled to access education, social services, and remote jobs. An estimated 42 million Americans do not have access to broadband today (including DSL and cable), and those with access often pay higher prices for lower speeds compared to customers in European and Asian countries. How to enhance the delivery of broadband infrastructure, especially for underserved and unserved places, is more crucial than ever. The US is behind other countries in addressing this digital divide.

Most research has focused on federal broadband policy, but in the US, states play a key role in providing public funds to ease some of the costs of expanding and upgrading infrastructure in underserved areas, and in easing regulatory barriers. Local leadership is also key, as localities are on the front line of closing the rural digital divide. Unable to wait for private market solutions, rural and low-income communities across the country are opting to work with local telephone and electric cooperatives, or run their own broadband networks.

This book highlights the role of state policy and innovative local response. We lift up the voices of state and local leaders in hopes their stories will provide insights for current federal and state broadband policy. We present a multi-level governance framework which emphasizes the role of local community planning and state policy. While recent federal policy dramatically increased support for broadband access (the CARES Act of 2020, the American Rescue Plan Act of 2021, the Infrastructure Investment and Jobs Act of 2021, which included the Broadband Equity, Access, and Deployment (BEAD) program), there is still much more work to be done. U.S. policy has expanded the definition of digital inclusion to encompass access, affordability, and adoption. While most of the chapters in this book focus on efforts to expand access via new infrastructure investment, digital inclusion also requires more attention to affordability and adoption if all people are to be included fully in society.

This book is based on several years of research, conducted from 2022–24, before the second Trump administration brought about additional policy changes. The book builds from a series of Cornell reports, Natassia Bravo's Ph.D. dissertation, and Edward Guo's senior thesis. A book is built on many collaborators. We wish to thank all the local and state leaders who agreed to be interviewed for the case studies. Your work is profiled in the case study Chapters 4–8. You have shown us the way to overcome market, financial, attitudinal, and policy barriers. We hope we have done justice to your innovation, leadership, and courage to build stronger communities.

We also wish to thank our funders: the Pew Charitable Trusts, the USDA National Institute of Food and Agriculture Program, and the USDA/Hatch Multi-State program administered by the Cornell Agricultural Experiment Station. Your commitment to digital inclusion and your special focus on rural communities made this book possible. We especially acknowledge the USDA National Institute of Food and Agriculture grant # 2021-67023-34437 for covering the costs to make this book open access. We hope this book will be widely used by practitioners and policymakers.

We wish to thank the students who helped conduct the interviews: Jane Bowman Brady, Melody Chen, Edward Guo, Shunyi Hu, Luke Kerr, Divine Maduakolam, Samuel Olafare Olagbaju, Shreni Rajbhandary, Ella Redmond, and Chen Wu. Your energy, enthusiasm, and insights were critical. Colleagues at the Pew Charitable Trusts provided close review of some of the material in Chapters 3, 5, and 6, especially Colby Humphrey and Kelly Wert, and the data check and fact check teams reviewed the earlier reports on which those chapters build. This is a difficult and important job, and we thank you. We would also like to thank those who reviewed earlier reports: Johannes Bauer, John Green, Colin Rhinesmith, and Revati Prasad. Your insights helped us see a bigger picture. A special thanks to Anna Read who invited us to begin this project years ago. Finally, we thank our editors at Routledge, Mark Brennan, Sarah Rae, and Selena Hosteler, for your faith in our team and your efforts to see this project through to publication.

We thank all our coauthors for their timely completion of chapters which integrate so beautifully into a cohesive narrative. A book is a major undertaking, but also a means to celebrate the innovative work happening across the country. We thank our families for tolerating the long days and late nights that were required to bring this book to fruition.

Now is the time for investment to close the digital divide. This book provides insights into how to do so, drawing from prior state and local policy experience. Much divides the US, but access to the Internet does not have to be the divider. It is up to us as a nation and as a society to recognize the need to bridge the digital divide and support those working so hard to do so. These are the community builders, the leaders, who help build a more inclusive society.

Mildred E. Warner, Natassia A. Bravo, and Duxixi (Ada) Shen
February 2025

ACRONYMS AND ABBREVIATIONS

ACP	–	Affordable Connectivity Program
APT	–	Alliance for Public Technology
ARPA	–	American Rescue Plan Act
ARRA	–	American Recovery and Reinvestment Act
BEAD	–	Broadband Equity Access and Development Program
BPL	–	Broadband over Power Line
BUD	–	Broadband Utility Districts (ME)
CAA	–	Consolidated Appropriations Act
CAF	–	Connect America Fund
CARES	–	Coronavirus Aid, Relief, and Economic Security Act
CCF	–	Community Capitals Framework
CCTHITA	–	Central Council of the Tlingit and Haida Indian Tribes of Alaska (AK)
CDC	–	Community Development Corporation
CETF	–	California Emerging Technology Fund
CNS	–	Cooperative Network Services
COE	–	County Office of Education (Santa Cruz, CA)
COW	–	Cell on Wheels
CTSB	–	Computer Science and Telecommunications Board
DBU	–	Downeast Broadband Utility (ME)
DEA	–	Digital Equity Act
DVI	–	Digital Volunteer Initiative
DMEA	–	Delta-Montrose Electric Association (CO)
DORA	–	Department of Regulatory Affairs (CO)
DOLA	–	Department of Local Affairs (CO)

DSL	–	Digital Subscriber Line
DVI	–	Digital Volunteer Initiative
EASC	–	Equal Access Santa Cruz (CA)
EBB	–	Emergency Broadband Benefit
EU	–	European Union
FCC	–	Federal Communications Commission
FTTH	–	fiber-to-the-home connections
FTTP	–	fiber-to-the-premise connections
GAO	–	Government Accountability Office
Gbps	–	Gigabits per second
HACNO	–	Housing Authority of the Choctaw Nation (OK)
HUD	–	US Department of Housing and Urban Development
IIJA	–	Infrastructure Investment and Jobs Act
IRRRB	–	Iron Range Resources and Rehabilitation Board (MN)
ISP	–	Internet Service Provider
LEAP	–	Lease to Purchase Program (Choctaw Nation)
LISC	–	Local Initiatives Support Corporation
Kbps	–	Kilobits per second
Mbps	–	Megabits per second
MCA	–	Maine Connectivity Authority
NAHASDA	–	Native American Housing Assistance and Self-Determination Act (Choctaw Nation)
NDEET	–	National Digital Education Extension Team
NESC	–	Northeast Service Cooperative
NOFO	–	Notices of Funding Opportunity
NTIA	–	National Telecommunications and Information Administration
OECD	–	Organization for Economic Cooperation and Development
PPP	–	Public-Private Partnership
PUB	–	Peninsula Utility for Broadband
PUC	–	Public Utilities Commission
RDOF	–	Rural Digital Opportunity Fund
RUS	–	Rural Utility Service
SDPBC	–	School District of Palm Beach County (FL)
SLA	–	Sustainable Livelihoods Approach
SRDC	–	Southern Rural Development Center
USCA	–	Universal Service Administrative Company
USDA	–	US Department of Agriculture
VATI	–	Virginia Technology Initiative (VA)
VoIP	–	Voice over Internet Protocol
WIA	–	Working Internet ASAP (ME)
YAB	–	Youth Advisory Board (Choctaw Nation)
YCEA	–	York County Economic Alliance (PA)

PART 1

Policy and Theoretical Background

We begin with a policy overview and theoretical framework. In Chapter 1, we explore the main factors in the digital divide and why it has persisted over time. We present a multi-level governance framework that shows the importance of federal policy, state policy, and local initiative. In Chapter 2, Johannes Bauer provides a history of telecommunication policy in the US. This chapter presents a dynamic, socio-technical framework of technology-society interaction to explain how digital equity is an outcome of contradictory technological, economic, and political factors. In Chapter 3, we review the experience with past state-level initiatives in 17 states and show how states were addressing the digital divide prior to the COVID-19 pandemic. We provide implications for states' addressing broader definitions of digital equity going forward. While Federal policy will change, the need to better understand how state, local and federal policy intersect will remain.

DOI: 10.4324/9781003619208-1

1

WHY DOES THE DIGITAL DIVIDE PERSIST IN THE US?

Mildred E. Warner and Natassia A. Bravo

* * *

I live on a small farm, just 9 miles from Cornell University, where I work as a professor of city and regional planning. The farm is lovely and the commute is under 20 minutes, but my house is in a digital desert. We didn't know we were buying a place off the grid when we bought it 35 years ago. We have electricity and land line telephone, but when the Internet became important in the 1990s, we quickly discovered we were off the grid and there would be little policy support to connect us.

Accessing the Internet has always been a challenge. Market and technology leave rural areas behind, and public policy has not prioritized Internet access, until now. DSL does not reach this far. A neighboring rural community put in a broadcast Internet tower with a local provider, but rolling hills and trees block a view from my house. We used HughesNet satellite for a while, but the latency meant I could not access key online functions, such as the Cornell library. So, we upgraded our cell phones and used them as hotspots for a few years. Each month we would burn through our data quickly. Often I would drive to the grocery store or library parking lot to download and upload emails. Libraries often design their Internet to cover the parking lot, so people without adequate service can access Internet, even when the library is closed. Then AT&T upgraded their tower from 3G to 5G and we were left in the dark again – even our phones no longer worked at home. 5G doesn't travel as far as 3G.

For years, cable was on the main highway (just a mile away), but they told us we would have to pay $17,000 to get them to come down our road! Then, a few years ago in 2018, Governor Cuomo threatened to kick out Comcast (now Spectrum) if they did not build out coverage as they had promised. The cable

DOI: 10.4324/9781003619208-2

company hired out-of-state crews to come up and string the lines. They strung cable on the roads below, behind, and above our road, but skipped us. We were going to be bypassed again. I called the chair of the county government board (as the county has some regulatory authority over cable) and asked if the cable company was going to finish the job and come down our road. She called the cable company and they came back and strung cable on our block. Good thing, because COVID hit the next year and I had to work online. It would have been impossible without cable. But other rural roads in the county are still not covered, and the crews are long gone.

If you live in a rural area, you are always left on the lagging end of technology and it is a constant struggle to get and stay connected. Right now, we use cable as our Internet source, thanks to the state pressuring the cable company to extend the lines. But it was just luck that our road was covered. Now the county is working to lay fiber – and it is going in on the road behind my house, which has more houses (our road has just four houses). This is great...for them, but my road will not be covered. The last mile, or half mile in this case, matters.

Accessing the Internet is a constant and unending struggle. I now realize my house will always be on the lagging end of technology. Cell phones no longer work here – thanks to the shorter range of 5G, so I had to switch to Internet phone. But cable could become a dying industry and then where will I be? Telephone offers a cautionary tale. Telephone is considered an essential service, but so few people use land lines anymore that the phone company doesn't regularly check the connector box a mile up the road. Mice build nests in the box, and about twice a year we lose phone service for a week or so. Access to the Internet is much more expensive than our land line, but Internet is not considered an essential service. Policy needs to catch up with reality. Internet access is critical for work, education, health care, shopping, and emergencies.

FIGURE 1.1   Catching Wi-Fi outside a public building. Photo by Mildred Warner.

I do not live in a remote rural area. But I fully understand the challenges of last mile service – and what it feels like to live on the other side of the digital divide (Figure 1.1).

* * *

The Role of Public Policy and Planning in Addressing the Digital Divide

The digital revolution has bypassed many Americans living on the margins, and universal access to high-speed, reliable, and affordable Internet remains an unrealized goal. The COVID-19 pandemic shed light on the decades-long and persistent digital divide, and this book showcases state and local efforts to address it. Ten years ago, the Federal Communications Commission (FCC) emphasized: "The need for broadband is everywhere, even if the business case is not" (FCC 15–25, 2015). Market solutions are not enough. Bridging the digital divide requires policy change, market regulation, technological innovation, and local planning and initiative. Consumers cannot do it alone. Unable to wait for private market solutions, unserved and underserved communities across the country are opting for a local approach – such as working with local telephone and electric cooperatives, or running their own broadband networks (Strover et al., 2021).

We need a community planning approach. Planning enables communities to identify needs, articulate their aspirations, and develop approaches to implement them. When the process is democratic and involves all stakeholders, it can facilitate dialogue to identify innovative solutions. Planning can repair the divides in society, but only if equity is centered (Reece, 2018; American Planning Association, 2019; Williams, 2024). Otherwise, planning can lead to capture by elite interests (Molotch, 1993). This is why community leadership and engagement are so critical.

Planning must also be supported by a regulatory and policy environment that centers digital equity and recognizes the need for universal access and affordability. Federal and state governments have stepped up to this challenge, but greater cooperation, coordination, and resource-sharing between institutions is needed. Addressing the digital divide requires a multi-level government approach, such as the one that we articulate in this book. No longer can the digital divide undermine our nation and our democracy. Now is the time to bridge that divide – and this requires money, shared power, and innovative planning, as shown by the cases in this book.

In contrast to other critical infrastructures where the federal government invested heavily to ensure broad access (water, telephone, electricity, highways), broadband has been left primarily to market forces. This has left behind geographic areas (rural) and populations (the poor, the elderly, and minorities) which are not profitable to serve. The COVID-19 pandemic brought the digital divide into stark relief, as school children could not access online lessons, and rural residents could not access telemedicine. This led to new investments by the federal government to enhance not only broadband access, but digital equity as well…at long last.

What do we know about prior policy efforts to address the digital divide? What can those efforts teach us about policy going forward? Despite the billions of dollars in federal subsidies to large telecommunications and cable companies and deregulation efforts to promote private investment, the country still has a long way to go to narrow the digital divide and achieve universal access. While the literature has mostly focused on federal broadband policy, this book explores the role of state policy, even before the pandemic, and creative local community responses. We showcase initiatives in rural communities and low-income urban ones, as those are the most left behind. We provide insights for policy design and local action at the community and regional levels.

At the onset of the pandemic, almost 15 million out of 50.7 million public-school students and 10% of teachers did not have adequate Internet access for distance learning (Chandra et al., 2020). Between 2019 and 2021, the number of individuals working from home rose from 9 million to 27.6 million (US Census Bureau, 2022). Telemedicine consultations rose dramatically, but were lower among households without high-speed Internet access (Kyle et al., 2021).

As of 2022, the FCC (2024) estimated that roughly 24 million Americans lacked access to Internet service that met the agency's speed benchmark. Broad-bandNow estimates the number is closer to 42 million (Cooper, 2024a). The pandemic has spurred the largest Federal investment in broadband to date. Mirroring historic Federal leadership in rural electrification and the creation of the interstate highway system, the Biden Administration allocated billions for infrastructure deployment under the American Rescue Plan Act of 2021 (ARPA) and the Infrastructure Investment and Jobs Act of 2021 (IIJA) (The White House, 2023). Several new funding streams for broadband deployment and digital equity were established, including the $42.45 billion Broadband Equity, Access and Deployment (BEAD) Program, the $2.75 billion Digital Equity Act (DEA) Program, and the $14.2 billion Affordable Connectivity Program (ACP). The incoming Trump administration will make changes to broadband policy, but the partnership with states, localities and industry will persist.

States have played a critical role in promoting broadband development in the past and will take a lead in the management and distribution of BEAD funds. This book provides insights on how state policy, including state grant eligibility criteria and restrictions on municipal broadband, impact local broadband initiatives. We show how local and regional initiatives navigate financial and regulatory constraints to expand access.

In this chapter, we describe the digital divide and present a multi-level governance framework for broadband, in which local governments and regional coalitions play an active role in broadband policymaking. This frames the chapters that follow, which profile state and local programs from across the United States.

Pandemic-Era Broadband Funding

The COVID-19 pandemic underscored the importance of government leadership when market solutions fall short, and triggered both unprecedented investment in broadband infrastructure and policy reform. While BEAD is the largest investment dedicated to broadband deployment to date, states and local governments were also able to harness billions in pandemic-relief funds from several economic stimulus bills – including the Coronavirus Aid, Relief, and Economic Security Act (CARES), the Consolidated Appropriations Act of 2021 (CAA), the American Rescue Plan Act of 2021 (ARPA) and the Infrastructure Investment and Jobs Act of 2021 (IIJA).

Several entities are involved in the administration of these funds. Chapter 2 provides an in-depth overview of the governmental role in broadband deployment, and how the US finally arrived at the policy window that now centers digital equity. Here, we briefly highlight two of the key actors in this field – the FCC and the National Telecommunications and Information Administration (NTIA).

The FCC was established by the Communications Act of 1934, and is in charge of regulating various types of communications services across the country. In 1997, the FCC created the Universal Service Fund, which receives contributions from telecommunications providers and Voice over Internet Protocol (VoIP) providers, and is home to several grant programs for rural deployment and Internet access for low-income households, schools and libraries, and healthcare providers. The FCC also plays an important role in setting the download and upload speeds benchmark for a service to qualify as "Broadband or high-speed Internet access." The amount of data transmitted over a network connection is measured in bits per second, and for almost a decade, the FCC's benchmark was 25 downloads and three upload Megabits per second. In 2024, it was upgraded to 100/20 Mbps. The FCC's benchmark has served as a reference for several federal and state broadband funding programs.

The NTIA, located within the Department of Commerce, provides grants for broadband deployment and adoption as well. The agency provided key aid to states through the State Broadband Initiative, launched with funds from the American Recovery and Reinvestment Act of 2009 (ARRA). This program provided funding to states for data collection and mapping efforts, and helped many launch their own broadband offices. The NTIA has also developed a Digital Inclusion Ecosystem that emphasizes all the agencies and efforts that must exist to truly address digital inclusion. Beyond access to high-capacity infrastructure, further measures must be taken to increase adoption and ensure affordability. The NTIA is also in charge of administering funds from BEAD and DEA.

We highlight relevant funding programs established or funded through recent economic stimulus bills in Appendix Table 1.1. Each of these funding streams brings its own guidelines, designed by the federal agency that oversees the funds. Conflict unfolds when federal and state authority overlap. The rules of federal and state programs can be mismatched, as each state designs its own funding policies. While states are expected to retain some autonomy over the distribution of BEAD-funded grants, they will need to comply with the NTIA's broader eligibility criteria and rules. In Chapter 3, we explore how states approached broadband funding in the years before the pandemic.

Digital equity: A Broader Notion of "Access"

Universal broadband access is not only a matter of infrastructure availability. Among the millions of Americans without access to high-speed and affordable Internet service, many live on the margins of society – including rural residents (Vogels, 2021a), individuals living in poverty (Vogels, 2021b), and minority groups (Atske and Perrin, 2021).

The digital divide has been partially attributed to historic underfunding in areas where low returns on investment do not justify the significant deployment costs (Turner Lee, 2024). The perception of low profitability gives providers little incentive to build or upgrade their infrastructure in rural, low-income, and high-minority urban neighborhoods, where these upgrades are less common (Yin and Sankin, 2022). "Digital redlining" has been an issue for years. Back in 2017, the National Digital Inclusion Alliance (NDIA) reported discriminatory deployment practices by AT&T that affected low-income neighborhoods in Cleveland, OH. While AT&T offered fiber-to-the-home in the county's suburbs and other markets in the US, the company had yet to upgrade their infrastructure in most of the city of Cleveland – especially high-poverty neighborhoods (Callahan, 2017). Since then, several cases of digital redlining have been identified (Leventoff, 2022). Low-income households also struggle to pay for Internet service, even where it is readily available.

Even where high-speed broadband Internet is available, many rural and low-income Americans do not have Internet subscriptions, lack access to a home computer, or the skills to use the Internet (Vogels, 2021a, 2021b; Turner Lee, 2024). Home Internet access subscriptions are lower among African-American and Hispanic adults (Atske and Perrin, 2021), as well as Native Americans (GAO, 2022).

A broader definition of access that considers the disproportionate impact of the digital divide on certain populations was needed. Thus, in 2021 federal policy expanded its focus on infrastructure to include digital equity. According to the NDIA, digital equity is achieved when all individuals have access to the technology, tools, and skills necessary to participate in society and the economy (NDIA, n.d.). The IIJA allocated $2.75 billion to the DEA of 2021, to support the development of state digital equity plans that identify and address barriers to Internet adoption and digital literacy. The ACP provided subsidies for Internet subscriptions to roughly 23 million households, but the program ran out of funds in May of 2024 and was not renewed by Congress (Hearn, 2024). Shifts in federal policy create challenges which state and local programs must navigate.

Challenges in Addressing the Digital Divide

The digital divide has many technical and market dimensions. These reflect economic and market forces, consumer education and demand, technology, industry structure, and the role of public policy.

Market Barriers

Lack of Competition and the Broadband Oligopoly

Some infrastructures, like water and sewer, are natural monopolies with economies of scale, and thus do not benefit from competition. This is *not* the case with broadband. Competition is beneficial, but the broadband market suffers from lack of it. Twenty-two percent of U.S. households with a broadband Internet subscription only have access to one provider (Leichtman Research Group, 2023). By the end of 2023, 96% of the broadband market was dominated by the largest cable and telecommunications companies (Light Reading, 2024). These include Comcast, Charter, AT&T Inc., Verizon, Altice/Optimum, Frontier, Lumen Technologies (formerly CenturyLink), Windstream, and Consolidated.

The challenges of building in rural areas were likely why the FCC first looked to large incumbents to close the digital divide. Under the direction of the FCC, the Universal Service Administrative Company (USAC) administers the Universal Service Fund, which is paid for with contributions from telecommunications providers. Back in 2015, USAC disbursed $10 billion to ten large and mid-size telecommunications companies. The logic was that 83% of unserved Americans lived in areas served by these carriers (Ali, 2020). These providers already owned critical infrastructure, and could easily expand into unserved areas. Following the same logic, electric cooperatives have moved into the broadband space. They already deploy fiber optic networks to upgrade their electric power grids into smart grids. They can lease excess fiber to rural Internet Service Providers (ISPs), or expand into last-mile service themselves (Read and Gong, 2022). However, public funding has not been as high for rural cooperatives as it has been for traditional telecoms. Until recently, many of these large incumbents continued to receive billions in federal subsidies to meet outdated speed requirements, while alternative providers – including utilities and cooperatives – had to compete for smaller pools of funding (Ali, 2020).

Some recipients of federal and state funding even began defaulting on their deployment commitments. Back in 2018, New York State threatened to kick out Charter Communications for this very reason (Brodkin, 2018). Recent federal grant recipients have requested that the FCC allow them to default on their commitments with limited penalties, claiming rising costs (Ferraro, 2024). This has critical implications. Because of these prior awards, such areas are generally not eligible for incoming federal funding from the BEAD Program (Rachfal et al., 2023). If these providers default on their commitments to build out coverage, the areas they had promised to cover will be left out of the new funding. They will be left behind both by the market and by policy. That hurts.

Corporate Capture

Corporate capture at the federal and state level has resulted in policymaking that protects industry interests and deters public competition (Crawford, 2020). For example, the largest ISPs reportedly spent more than $234 million on lobbying during the 116th United States Congress (2019–2020). They fought against net neutrality, data transparency on prices and actual speeds, municipal broadband, and efforts to prevent service terminations during the pandemic. They advocated for "technology neutrality" – meaning, for no broadband technology to be prioritized over another – and lower service speed requirements to ensure they remained eligible for public funding (Common Cause, 2021). Incumbent ISPs also have been criticized for locking areas out of federal funding (Brodkin, 2020), and blocking applications from competitors (Muller, 2022), including municipal broadband initiatives (Block, 2022). Net neutrality, which was repealed in 2017, treated broadband Internet service as a utility and prohibited discriminatory practices by ISPs. It was restored in early 2024, as the FCC declared broadband Internet an essential service (Teale, 2024). Nonetheless, the industry was unwilling to build or upgrade their infrastructure in less profitable areas, and simultaneously opposed the allocation of public dollars going to municipal broadband efforts.

Does this song sound familiar? Lest we forget – AT&T Inc., Verizon, and Lumen are the last of the remaining "Baby Bells." In 1974, the U.S. Department of Justice brought an antitrust lawsuit against AT&T's predecessor, the American Telephone and Telegraph Company (AT&T Company). The AT&T Company was the successor of the Bell Telephone Company, founded in 1877 and holder of Alexander Graham Bell's telephone patent. In 1982, the Company agreed to divest from its system of local telephone companies (the "Bell System"), which was then re-organized into seven independent companies (the "Baby Bells"). Before its break-up, the Bell System served 81% of all telephones in the US

(Hart et al., 1982). AT&T essentially controlled local and long-distance calling services, as well as the manufacturing of telephone and network equipment (Sullivan and Hertz, 1990). Early on, AT&T began acquiring its competitors; later, it allowed competitors to hold regional monopolies in areas outside the Bell System. This served to discourage competitors from expanding, and the government from suing AT&T for anticompetitive behavior (Thierer, 1994).

According to Thierer (1994), misguided federal and state policies facilitated AT&T's domination as well. The notion of telephone service as a natural monopoly gained popularity among policymakers, who began discouraging competition. Independent telephone companies were barred from building new lines in areas that were already "served." The Bell System was allowed to charge discriminatory rates to cross-subsidize its expansion, gaining advantage over local competitors. With broadband, the high costs of entry already tilt the balance in favor of large telecommunications and cable companies. While public broadband funding could level the playing field, new entrants are unlikely to secure federal or state funding for communities that already have at least one provider. Limited competition is unlikely to be solved with public funding, and corporate capture ensures that policy will prioritize market interests over digital equity.

Technology Barriers

Broadband access in the US has been described as both a market failure (Edwards, 2009) and a policy failure (Ali, 2021). The distribution of billions in subsidies has been complicated by inaccurate broadband coverage maps (Ali, 2021), which help determine which areas are "unserved" and thus qualify for public funding. The FCC's Fixed Broadband Deployment Maps were criticized for overestimating rural broadband access (Barrett and Arseneau, 2022). The data was collected at the census block level, and submitted by ISPs themselves. An entire census block could be designated as "served" if a single house was served, or could be served within a short timeframe. In addition, providers could report advertised – as opposed to actual – service speeds.

There is also the issue of defining which download and upload speeds and technologies qualify as "broadband." Since the Telecommunications Act of 1996, the FCC has undertaken this task. The agency has long maintained a policy of "technology neutrality" – allowing a broad range of wired and wireless technologies to be equally eligible for funding, as long as they meet the agency's benchmark. However, technology neutrality raises some challenges, as we learn next.

Broadband Technologies Are Not Interchangeable

Broadband technologies are vastly different in terms of their data transmission speeds, capacity, and reliability. Fiber optic networks are the gold standard of broadband technologies, as they can transport large amounts of data at exceptionally high speeds. These networks use light signals that travel across strands of glass, whereas cable Internet and Digital Subscriber Line (DSL) use electric signals instead. They are carried by coaxial cables or legacy copper phone landlines, which are more widely available (CenturyLink, n.d.). DSL is the oldest and least expensive option, but it is also significantly slower than fiber optic and cable (Cooper, 2023b). Overall, fiber optic is considered a sensible long-term investment due to its reliability, speed, and capacity to accommodate growing user traffic (Afflerbach, 2022).

One might wonder why fiber optic is not the universal choice for new infrastructure. For one, there are the high costs of deployment. Broadband infrastructure is a capital-intensive investment that involves various construction, financing, operational, and permitting costs. In particular, the cost of fiber optic deployment can rise due to several factors – costs of labor, materials and network equipment, and type of installation (Kim, 2024). Depending on the terrain, providers might opt for an aerial or underground installation of fiber conduit. Both are costly. An underground installation requires excavation to bury the conduit in a trench. An aerial installation requires pole attachments, which involves seeking permits from pole owners and other ISPs that also own conduit attached to the same pole. In rural areas, limited staff and high rental rates can cause delays in pole attachments (Hardesty, 2023).

While federal or state broadband grants can help offset a percentage of construction costs, they may still fall short as incentives to serve remote, low-density, and high-poverty areas. Where fiber-to-the-home is not a viable option, wireless technologies like fixed wireless and satellite Internet are proposed as acceptable alternatives. They are easier or less costly to deploy than fiber optic broadband – but are not without their own challenges. Fixed wireless access can be slowed down or disrupted by high network congestion, obstructed lines-of-sight between cell towers and extreme weather conditions (TP-Link, 2023). Satellite Internet is plagued by similar issues, and is both slower and more expensive than cable (Cooper, 2024b). Next-generation satellite technologies are now promising to deliver higher speeds, in the hopes of remaining eligible for funding (Baumgartner, 2023). Indeed, satellite provider Starlink is emerging as an important player, but concerns remain around cost, speed and reliability. Fiber is the best long term investment.

The Pitfalls of Technology Neutrality

Policy is another reason why fiber is not yet a universal choice. The broadband industry has a vested interest in keeping federal and state broadband definitions as broad as possible, so different technologies can qualify for funding. However, we should remember that federal and state deployment grants are limited. If grants are going to DSL and satellite, then there will be less funding available for fiber optic. Despite a significant gap between technologies, older and cheaper technologies were long considered "acceptable" options for rural areas.

For years, federal and state programs were reluctant to prioritize fiber openly, exclude older technologies or raise their download and upload speed thresholds. By keeping their eligibility requirements broad and flexible, these programs hoped to raise interest in serving rural or high-poverty areas. However, as their requirements failed to keep up to date with technology changes until recently, federal dollars were spent on broadband projects that no longer meet the FCC's current download/upload speeds threshold (100/20 Mbps). Some projects deliver speeds as low as 10/1 Mbps, which can be achieved with older technologies like DSL or satellite (Neenan, 2023). Meanwhile, many wealthy urban and suburban areas already have access to fiber, which delivers speeds of 1 Gigabit (1,000/1,000 Mbps) or higher – well above the FCC's benchmark.

Even if the technology is outdated, these areas are now considered "served." Federal agencies and state broadband programs are cautious about subsidizing construction in "served" areas, which is known as "overbuilding." Once again, we are reminded that broadband technologies are not interchangeable, and there is little incentive for providers to deploy the latest technologies in rural communities and low-income urban neighborhoods. This becomes an issue when slower and less stable technologies, or that impose data caps, are unable to manage growing user traffic or applications that consume large amounts of data (bits per second). Some of these applications have become part of everyday life, such as video streaming and video conferencing services. For example, Netflix recommends between 3 and 15 Mbps for high-definition video streaming – for a single device. During the pandemic, households needed a network connection with sufficient capacity to support multiple users consuming large amounts of data at the same time. Without such capacity, individual connections slow down and become less stable, and so entire families lack the tools to participate in remote work and virtual learning.

The unsurprising outcome is that the digital divide persists, as illustrated in the story at the beginning of this chapter. Federal and state policies must ensure that public funds are used for broadband infrastructure and upgrades that will stand the test of time. Otherwise, the technology gap between competitive and captured or ignored markets will continue to widen.

Rural Broadband Projects Involve More Costs

Finally, rural communities have different infrastructure needs than their urban counterparts. Owing to their remoteness, low-population density, and difficult terrain, rural communities often struggle with access both to last-mile and middle-mile networks, as described in Mildred's personal story above. Most of us are familiar with last-mile networks, which serve homes and businesses directly. To gain access to the broader Internet, these local networks must plug into a middle-mile network, which often operate at the county, regional, or state level

(Arnold and Sallet 2020). Middle-mile networks carry large amounts of data at great speeds and over long distances, and also connect government buildings and community anchor institutions – including libraries, hospitals, and schools. For remote rural communities that do not have access to a middle-mile network nearby, providers must factor-in the cost of building the middle-mile segment as part of their last-mile projects (Perrin, 2024). This is not an issue in urban areas, where the demand justifies this investment and there is a greater supply of middle-mile networks (King, 2024).

Rural broadband projects tend to be more expensive because some investment in middle-mile infrastructure is needed. The cost could impact a provider's decision to serve remote communities, driving them to seek urban-adjacent markets instead. Public funding acts as a critical incentive for rural deployment, but funding programs largely focus on the last mile. For example, note the contrast between the $42.45 billion set aside for BEAD and the $1 billion for the Enable Middle Mile Program (Whipple, 2023).

The digital divide is a decades-long issue, and there has been a constant flow of public dollars for broadband infrastructure circulating through federal and state broadband programs. However, billions in grants and loans can only go so far when systematic issues like digital discrimination and corporate capture go unacknowledged. Digital discrimination has resulted in outdated infrastructure for rural and marginalized populations. Corporate capture resulted in lower broadband service standards and halted competition in monopoly areas. Outdated broadband definitions resulted in billions of dollars going to infrastructure that quickly became obsolete. These issues became more salient during the COVID-19 pandemic, when various essential activities were disrupted by the pandemic and had to shift to a virtual mode.

Federal and state broadband programs can play a key role in achieving universal access and addressing the barriers to digital equity. This book focuses on the leadership of local actors and institutions in broadband expansion, and how state and federal policy supports their efforts. Chapter 2 provides background on federal policy, and Chapter 3 explores the role of state broadband funding before the pandemic.

Multi-Level Governance Framework

Why has the digital divide persisted, despite policy efforts to close it for more than four decades? We provide a critical lens to understand this failure, and we emphasize the importance of a multi-level governance framework. We call for centering public values in planning as we focus on state broadband policy designs and the implications for local action.

Access to community assets is critical to center public values in broadband planning and deployment. Communities need leverage that allows them to play a more active role and retain some control in these partnerships, to ensure that public objectives (service quality and reliability, affordability, and universal access) are met. While broadband is not a public utility or a natural monopoly, it has become an essential service. The ideal public infrastructure should be universally available and accessible to all individuals, interlinked with other essential services, and support positive externalities for the community (O'Neill, 2010).

Infrastructure, like broadband, requires a multi-level governance framework. Homsy, Liu, and Warner (2019) emphasize the importance of coordinated state and federal action, and the critical role of local action – for knowledge sharing and support of local initiatives. The COVID-19 pandemic underscored the importance of government leadership where market solutions fall short, and triggered both unprecedented investment in broadband infrastructure and policy reform. The Biden Administration made clear that it wanted the new federal funding to promote transformative change and address the structural divides made so clear during the pandemic (Warner et al., 2023). Communities stepped up and used the funds to address critical infrastructure and service needs – and equity was often embedded in project design (Diaz Torres and Warner, 2024). These funds were part of an intentional "politics of repair" pursued by the Biden Administration to address the dysfunction in the broken fiscal federalism of prior administrations (Xu and Warner, 2024).

The broadband literature has primarily focused on federal funding strategies, but state broadband programs will play a critical role in the distribution of BEAD awards. The federal-state dynamic of overlapping rules means that states do not have full discretion over how to address the digital divide. Chapter 3 describes trends in state broadband funding before the pandemic, which will have implications for future federal broadband policy. Local communities play a critical role, which we highlight in Chapters 4–9 of this book.

Figure 1.2 provides a schematic of our multi-level governance framework. At each level, there are policy issues to address, technical factors, and market players. At the federal level, we have outlined key challenges – most subsidies have gone to large providers who do not address digital equity, and incorrect digital coverage maps prevent communities from gaining access to funding. At the state level, we explore the impact of strict grant requirements which limit local action. These include: local match requirements, municipal broadband restrictions, and restrictions on blending state and federal funds. There are a range of technical issues (speed, technology), and organizational issues (middle mile, anchor institutions) which must be addressed. The local level is where it all must come together, and this book explores how local leadership gathers resources and promotes strategic action with a range of partners from the public and private sectors.

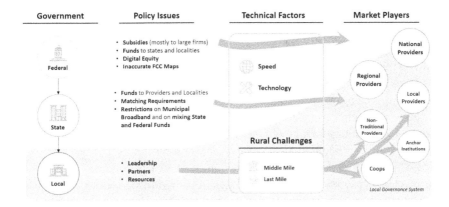

| Government | Policy Issues | Technical Factors | Market Players |

FIGURE 1.2   Multi-level governance framework for broadband. Image by authors.

While broadband is a local matter, policy is enacted at the federal and state level. The federal-level designs the broader standards for broadband, and states can adopt more detailed regulation that fits their unique challenges and needs. At the local level, communities proactively seek private investment or invest in their own publicly owned broadband infrastructure. Conflict emerges where there is a mismatch of priorities. Corporate capture may prioritize industry interests in legislation and program design. When local initiatives (such as publicly owned and/or publicly operated broadband networks) threaten industry interests, they can be restricted. Policy changes are driven by this power struggle between the industry, federal agencies, and states. This shapes how local actors (local governments, broadband advocates, and small providers) can address the digital divide. Broadband is a local problem, but the local response is limited by regulation and lack of resources.

Beyond coordinating, cooperating, and sharing broadband funding with local governments, local actors must be empowered to participate in, and influence policy design. Local officials, broadband advocates, and technical experts have first-hand knowledge about community connectivity needs, but their role is often limited to planning. What's more, there is little local input over state regulation that could prevent communities from owning or operating their own broadband networks, or over grant rules that may disqualify communities that are still underserved. Parts 2 and 3 of this book showcase the leadership of local officials, broadband advocates, and rural and urban providers as they navigate financial and regulatory challenges.

The local scope of action is largely defined by states, which are not always aligned with federal goals. At least sixteen states still preempt – to varying degrees – the ability of local governments to own, build, and/or operate and maintain a broadband network (Cooper, 2023a). Many of these municipal broadband restrictions have been promoted by the industry (Holmes, 2014). Municipal broadband projects in these states are discouraged or outright precluded from competing for state funding. This has implications for the distribution of BEAD funds, as municipal networks are now eligible for funding but restricted by state law (Ali et al., 2024). We showcase in Chapter 7 how localities, in preempting states, work around these restrictions.

State regulation can hinder local leadership, leaving unserved and underserved communities with no options other than private investment. Without the threat of competition from local governments or electric cooperatives, there will be no incentive for providers to enter an unserved area, or upgrade their services or reach every home in areas where they already provide service. Even if local governments outsource their network operation and lease their infrastructure to providers, retaining public ownership can be critical. Public ownership can give localities leverage in these contracts, and guarantee local input over how broadband projects are designed, planned, and managed. Chapters 5 and 6 give specific attention to how state policies impact rural broadband initiatives in Minnesota, Colorado, and Maine.

In Chapters 4–9, we learn how local officials, planners, providers, and advocates leverage community resources, and state and federal funding to improve Internet access in underserved areas. Chapter 4 presents a community resilience framework, in which communities build on existing resources and active agents to promote strategic action for equity and impact (Magis, 2010). In a dynamic broadband environment characterized by change and uncertainty, resilient communities intentionally develop a collective capacity to respond to and influence change, and develop new trajectories for the community's future.

Outline of Book

Processes of infrastructural change are driven by historic, economic, regulatory, and political forces. Uneven access to infrastructure creates regions of inequality. This book explores the role of two factors in addressing the digital divide – broadband policy and local initiative. Federal policy since the COVID-19 pandemic has centered public concerns like universal access, affordability and inclusion in policy design and network planning. Federal agencies and states are in charge of regulating broadband delivery and designing grant funding rules, but they must work together with broadband providers and local leaders to navigate regulatory and financial challenges. Previous studies have primarily focused on the impact of federal agencies and national telecom providers. We highlight the critical role of states and local governments in centering public values in broadband delivery.

This book has four parts. Part 1 provides an overview of the forces driving the digital divide, the key actors involved, and the impacts of federal and state policy. In this chapter, we have provided a policy and theoretical framework. In Chapter 2, Johannes Bauer explores how federal policy has evolved and lays out a framework for addressing digital equity as we tackle future challenges. In Chapter 3, we examine how state-level eligibility criteria and regulation impact the distribution of state broadband grants. We use local-level demographic, socioeconomic, and market characteristics, to assess whether state funds reach counties that are disproportionately impacted by the digital divide.

Part 2 explores the role of local action. In Chapter 4, we articulate a community resilience framework for local action, illustrating this with cases from Brownsville, Texas and Santa Cruz, California. Chapter 5 provides examples from Minnesota, an early state leader in addressing rural broadband. Chapter 6 presents cases from Colorado and Maine, states recognized for taking a regional approach to addressing the digital divide. Chapter 7 explores how localities navigate state preemption of municipal broadband by building partnerships with providers.

Part 3 explores the role of institutional leadership for digital equity. In this section, we showcase the role of Indigenous leadership in addressing the digital divide in two cases from the Choctaw Nation of Oklahoma, and native Alaskan communities (Chapter 8). In Chapter 9, John Green, Roberto Gallardo, and Roseanne Scammahorn explore the role of Cooperative Extension in addressing the multiple dimensions of the digital divide across Southern states in the US.

Part 4 discusses how policy and planning in a multi-level governance system can be reformed to center public values. Chapter 10 summarizes the role of state and federal policy to address digital equity, and how communities address digital equity concerns.

This book showcases examples of how broadband infrastructure is designed and planned, how localities cooperate with local advocates and regional providers, and the challenges that shape this process – including state regulation, costs of deployment, geography, and government capacity. The book identifies the challenges for local broadband initiatives related to state regulation and federal-state overlaps of authority, and makes recommendations for policy reform. We hope this will help guide future policy and planning.

Bibliography

Afflerbach, A. (2022). Fixed Wireless Technologies and Their Suitability for Broadband Delivery [Report]. Benton Institute for Broadband & Society. https://www.benton.org/sites/default/files/FixedWireless.pdf

Ali, C. (2020). The Politics of Good Enough: Rural Broadband and Policy Failure in the United States. *International Journal of Communication* 14, 5982–6004.

Ali, C. (2021). *Farm Fresh Broadband: The Politics of Rural Connectivity*. Information Policy Series. Cambridge, MA: The MIT Press. 306. https://doi.org/10.7551/mitpress/12822.001.0001

Ali, C., Berman, D. E., Forde, S. L., Meinrath, S. and Pickard, V. (2024, May 16). The Bad Business of BEAD. Benton Institute for Broadband & Society. https://www.benton.org/blog/bad-business-bead

American Planning Association. (2019). *Planning for Equity Policy Guide*. https://planning-org-uploaded-media.s3.amazonaws.com/publication/download_pdf/Planning-for-Equity-Policy-Guide-rev.pdfr

Arnold, J. and Sallet, J. (December 1, 2020). If We Build Them, Will They Come? Lessons From Open-Access, Middle-Mile Networks [Report]. Benton Institute for Broadband & Society. https://www.benton.org/publications/middle-mile

Atske, S. and Perrin, A. (2021). Home Broadband Adoption, Computer Ownership Vary by Race, Ethnicity in the U.S. Pew Research Center. https://www.pewresearch.org/short-reads/2021/07/16/home-broadband-adoption-computer-ownership-vary-by-race-ethnicity-in-the-u-s/

Barrett, R. and Arseneau, K. (2022, January 7). With Poor Data, Deficient Requirements and Little Oversight, Massive Public Spending Still Hasn't Solved the Rural Internet Access Problem. *Milwaukee Journal Sentinel.* https://www.jsonline.com/in-depth/news/2021/07/14/weve-spent-billions-provide-broadband-rural-areas-what-failed-wisconsin/7145014002/

Baumgartner, J. (2023, May 4). Viasat Sizes Up BEAD Opportunity. *Light Reading.* https://www.lightreading.com/digital-divide/viasat-sizes-up-bead-opportunity

Block, B. (2022, March 30). How Comcast and Other Telecoms Scuttle Rural WA Broadband Efforts. Cascade Public Media. https://crosscut.com/news/2022/03/how-comcast-and-other-telecoms-scuttle-rural-wa-broadband-efforts

Brodkin, J. (2018, July 26). NY Threatens to Kick Charter Out of the State after Broadband Failures. *ArsTechnica.* https://arstechnica.com/tech-policy/2018/07/ny-threatens-to-kick-charter-out-of-the-state-after-broadband-failures/https://arstechnica.com/tech-policy/2018/07/ny-threatens-to-kick-charter-out-of-the-state-after-broadband-failures/

Brodkin, J. (2020, May 1). Frontier, Amid Bankruptcy, Is Suspected of Lying about Broadband Expansion. *ArsTechnica.* https://arstechnica.com/tech-policy/2020/05/frontier-amid-bankruptcy-is-suspected-of-lying-about-broadband-expansion/

Callahan. (2017, March 10). AT&T's Digital Redlining of Cleveland. NDIA. https://www.digitalinclusion.org/blog/2017/03/10/atts-digital-redlining-of-cleveland/

Chandra, S., Chang, A., Day, L., Fazlullah, A., Liu, J., McBride, L., Mudalige, T., Weiss, D., (2020). Closing the K–12 Digital Divide in the Age of Distance Learning. San Francisco, CA: Common Sense Media. Boston, Massachusetts, Boston Consulting Group. https://www.commonsensemedia.org/sites/default/files/featured-content/files/common_sense_media_report_final_7_1_3pm_web.pdf

Common Cause. (2021). Broadband Gatekeepers: How ISP Lobbying and Political Influence Shapes the Digital Divide. https://www.commoncause.org/wp-content/uploads/2021/07/CCBroadbandGatekeepers_WEB1.pdf

Congressional Research Service. (2021a). Overview of the Universal Service Fund and Selected Federal Broadband Programs [Report]. https://www.everycrsreport.com/files/2021-04-30_R46780_9d4a26447ef4ce55811cfb5b35249e82c4552ac7.pdf

Congressional Research Service. (2021b). The Consolidated Appropriations Act, 2021 Broadband Provisions: In Brief [Report]. https://crsreports.congress.gov/product/pdf/R/R46701

Congressional Research Service. (2021c). Infrastructure Investment and Jobs Act: Funding for USDA Rural Broadband Programs. https://crsreports.congress.gov/product/pdf/IF/IF11918

Cooper, T. (2024a, April 9). Broadband Availability Is Overstated in Every State. *BroadbandNow*. https://broadbandnow.com/research/broadband-overstated-in-every-state

Cooper, T. (2024b, April 3). Satellite Internet: Pros, Cons and Terminology. *BroadbandNow*. https://broadbandnow.com/guides/satellite-internet-pros-and-cons

Cooper, T. (2023a, November 17). Municipal Broadband 2023: 16 States Still Restrict Community Broadband. *BroadbandNow*. https://broadbandnow.com/report/municipal-broadband-roadblocks

Cooper, T. (2023b, October 26). DSL vs. Cable vs. Fiber: Which Internet Option Is the Best? *BroadbandNow*. https://broadbandnow.com/guides/dsl-vs-cable-vs-fiber

Council of State Governments. (n.d.). Infrastructure Investment and Jobs Act: Broadband Affordability and Infrastructure. https://web.csg.org/recovery/wp-content/uploads/sites/24/2021/11/Infrastructure-Investment-and-Jobs-Act.pdf

Crawford, S. (2014) Captive Audience: The Telecom Industry and Monopoly Power in the New Gilded Age. New Haven, CT: Yale University Press.

Diaz Torres, P. and M. E. Warner (2024). A Policy Window for Equity? The American Rescue Plan and Local Government Response, *Journal of Urban Affairs*, forthcoming. https://doi.org/10.1080/07352166.2024.2365788

Edwards, J. C. (2009). Digital Deliverance: Dragging Rural America, Kicking and Screaming, into the Information Economy. Lanham, MD: University Press of America.

Federal Communications Commission. (2024, March 14). 2024 Section 706 Report. https://docs.fcc.gov/public/attachments/FCC-24-27A1.pdf

Federal Communications Commission. (2015, March 12). Memorandum Opinion and Order: City of Wilson, North Carolina Petition for Preemption of North Carolina General Statute Sections 160A-340 et seq.; The Electric Power Board of Chattanooga, Tennessee Petition for Preemption of a Portion of Tennessee Code Annotated Section 7–52–601. WC Docket Nos. 14–115 and 14–116. https://docs.fcc.gov/public/attachments/FCC-15-25A1_Rcd.pdf

Ferraro, N. (2024). Broadband Coalition Asks FCC to Grant RDOF Relief for BEAD's Sake. *Light Reading*. https://www.lightreading.com/broadband/broadband-coalition-asks-fcc-to-grant-rdof-relief-for-bead-s-sake

GAO (2022). Tribal Broadband: National Strategy and Coordination Framework Needed to Increase Access. https://www.gao.gov/products/gao-22-104421

Hardesty, L. (2023). FCC still hasn't ruled on pole attachments, presenting a hiccup to aerial fiber. *Fierce Network*. https://www.fierce-network.com/broadband/fcc-still-hasnt-ruled-pole-attachments-presenting-hiccup-aerial-fiber

Hart, B. A., Nave, A. M., Raskob, A. W. Jr. and Thomason, J. C. (1982). Telephone Areas Serviced by Bell and Independent Companies in the United States [Report]. NTIA. https://its.ntia.gov/umbraco/surface/download/publication?reportNumber=82-97_ocr.pdf

Hearn, T. (2024, May 31). White House Laments Passing of the ACP. Broadband Breakfast. https://broadbandbreakfast.com/white-house-laments-passing-of-the-acp/

Holmes, A. (2014, August 28). How Big Telecom Smothers City-Run Broadband. The Center for Public Integrity. https://publicintegrity.org/inequality-poverty-opportunity/how-big-telecom-smothers-city-run-broadband/

Homsy, G., Liu, Z. and Warner, M. E. (2019). Multilevel Governance: Framing the Integration of Top-Down and Bottom-Up Policymaking. *International Journal of Public Administration*, 42(7): 572–582. https://doi.org/10.1080/01900692.2018.1491597

Internet for All. (n.d.). BEAD Allocation Methodology. Retrieved on June 3, 2024 from https://www.internetforall.gov/program/broadband-equity-access-and-deployment-bead-program/bead-allocation-methodology

Kim, J. (2024, January 15) "Fiber Optic Network Construction: Process and Build Costs". Dgtl Infra. https://dgtlinfra.com/fiber-optic-network-construction-process-costs/

King, J. (2024, May 16). Heed the Middle Mile for Rural Broadband, Industry Leaders Caution. *Fierce Network*. https://www.fierce-network.com/broadband/heed-middle-mile-industry-leaders-caution

Kyle, M. A., Blendon, R. J., Findling, M. G. and Benson, J. M. (2021). Telehealth Use and Satisfaction among U.S. Households: Results of a National Survey. *Journal of Patient Experience*, 2021: 8. https://doi.org/10.1177/23743735211052737

Leichtman Research Group. (2023, December 11). 92% of U.S. Households Get an Internet Service at Home. https://leichtmanresearch.com/wp-content/uploads/2023/12/LRG-Press-Release-12-11-2023.pdf

Leventoff, J. (2022, June 10). Digital Redlining: Why Some Older Adults Overpay for Bad Internet. National Council on Aging. https://www.ncoa.org/article/digital-redlining-why-some-older-adults-overpay-for-bad-internet

Light Reading (2024, March 7). About 3,500,000 Added Broadband from Top Providers in 2023 – Leichtman Research Group. https://www.lightreading.com/broadband/about-3-5m-added-broadband-from-top-providers-in-2023-leichtman-research-group

Magis, K. (2010). Community resilience: An indicator of social sustainability. *Society & Natural Resources*, 23(5), 401–416. https://doi.org/10.1080/08941920903305674

Molotch, H. (1993). The Political Economy of Growth Machines. *Journal of Urban Affairs*, 15(1): 29–53. https://doi.org/10.1111/j.1467-9906.1993.tb00301.x

Muller, W. (2022, September 1). Telecom Giant Moves to Stop Broadband Grant for Northeast Louisiana. Louisiana Illuminator. https://lailluminator.com/2022/09/01/telecom-giant-moves-to-stop-federal-broadband-grant-for-northeast-louisiana/

Neenan, J. (2023, November 9). 'It Was Graft': How the FCC's CAF II Program Became a Money Sink. *Broadband Breakfast*. https://broadbandbreakfast.com/it-was-graft-how-the-fccs-caf-ii-program-became-a-money-sink/

NDIA. (n.d.). NDIA Definitions. https://www.digitalinclusion.org/definitions/

NTIA. (2022). Notice of Funding Opportunity. Broadband Equity, Access and Deployment Program. https://broadbandusa.ntia.doc.gov/sites/default/files/2022-05/BEAD%20NOFO.pdf

O'Neill, P. (2010). Infrastructure Financing and Operation in the Contemporary City. *Geographic Research*, 48(1): 3–12. https://doi.org/10.1111/j.1745-5871.2009.00606.x

Perrin, S. (2024, April 18). Nothing Is Middling about the Middle Mile. *Light Reading*. https://www.lightreading.com/optical-networking/nothing-is-middling-about-the-middle-mile

Rachfal, C. L., Benson, L. S. and Zhu, L. (2023). Federal Funding for Broadband Deployment: Agencies and Considerations for Congress [Report]. Congressional Research Service. https://crsreports.congress.gov/product/pdf/R/R47883

Read, A. and Gong, L. (2022, March 29). States Considering Range of Options to Bring Broadband to Rural America. The Pew Charitable Trusts. https://www.pewtrusts.org/en/research-and-analysis/articles/2022/03/29/states-considering-range-of-options-to-bring-broadband-to-rural-america

Read, A. and Wert, K. (2021, December 6). How States Are Using Pandemic Relief Funds to Boost Broadband Access. The Pew Charitable Trusts. https://www.pewtrusts.org/en/research-and-analysis/articles/2021/12/06/how-states-are-using-pandemic-relief-funds-to-boost-broadband-access

Reece, J. W. (2018). In Pursuit of a Twenty-First Century Just City: The Evolution of Equity Planning Theory and Practice. *Journal of Planning Literature*, 33(3): 299–309. https://doi.org/10.1177/0885412218754519

Sullivan, L. A. and Hertz, E. (1990). The AT&T Antitrust Consent Decree: Should Congress Change the Rules? *High Technology Law Journal*, 5(2): 233–255. https://www.jstor.org/stable/pdf/24122356.pdf

Strover, S., Riedl, M. J. and Dickey, S. (2021). Scoping New Policy Frameworks for Local and Community Broadband Networks. *Telecommunications Policy*, 45(10): 1–13. https://dx.doi.org/10.2139/ssrn.3427571

Teale, C. (2024, April 25). FCC Reinstates Net Neutrality Rules. *Route Fifty*. https://www.route-fifty.com/infrastructure/2024/04/fcc-reinstates-net-neutrality-rules/396079/

The Pew Charitable Trusts. (2020, November 16). States Tap Federal CARES Act to Expand Broadband. https://www.pewtrusts.org/en/research-and-analysis/issue-briefs/2020/11/states-tap-federal-cares-act-to-expand-broadband

The White House (2023, June 26) Remarks by President Biden on Broadband Investments. https://bidenwhitehouse.archives.gov/briefing-room/speeches-remarks/2023/06/26/remarks-by-president-biden-on-broadband-investments/

Thierer, A. (1994). Unnatural Monopoly: Critical Moments in the Development of the Bell System Monopoly. *CATO Journal*, 14(2): 267–286. https://www.cato.org/sites/cato.org/files/serials/files/cato-journal/1994/11/cj14n2-6.pdf

TP-Link. (2023, March 23). Fixed Wireless Access: Obstacles and Regional Availability. https://community.tp-link.com/us/home/forum/topic/603786

Turner Lee, N. (2024). *Digitally Invisible: How the Internet Is Creating the New Underclass*. Brookings.

US Census Bureau (2022) The Number of People Primarily Working From Home Tripled Between 2019 and 2021. https://www.census.gov/newsroom/press-releases/2022/people-working-from-home.html

Varn, J., Gong, L. and Humphrey, C. (2023). How State Broadband Offices Are Using Initial Dollars from Capital Projects Fund. The Pew Charitable Trusts. Available online: https://www.pewtrusts.org/en/research-and-analysis/articles/2023/05/23/how-state-broadband-offices-are-using-initial-dollars-from-capital-projects-fund

Vogels, E. (2021a). Some Digital Divides Persist between Rural, Urban and Suburban America. Pew Research Center. https://www.pewresearch.org/short-reads/2021/08/19/some-digital-divides-persist-between-rural-urban-and-suburban-america/

Vogels, E. (2021b). Digital Divide Persists Even as American with Lower Incomes Make Gains in Tech Adoption. Pew Research Center. https://www.pewresearch.org/short-reads/2021/06/22/digital-divide-persists-even-as-americans-with-lower-incomes-make-gains-in-tech-adoption/

Warner, M. E., Kelly, P. M. and Zhang, X. (2023). Challenging Austerity under the Covid-19 State. *Cambridge Journal of Regions, Economy and Society*, 16(1): 197–209. https://doi.org/10.1093/cjres/rsac032

Whipple, T. (2023, May 17). But for the BEAD Program, $1 Billion in Middle Mile Funding Wouldn't Be Enough. *Broadband Breakfast*. https://broadbandbreakfast.com/but-for-the-bead-program-1-billion-in-middle-mile-funding-wouldnt-be-enough/

Williams, R. A. (2024). From Racial to Reparative Planning: Confronting the White Side of Planning. *Journal of Planning Education and Research*, 44(1): 64–74. https://doi.org/10.1177/0739456X20946416

Xu, Y. and Warner, M. E. (2024). Fiscal Federalism, ARPA and the Politics of Repair, *Publius: The Journal of Federalism*, 54(3): 487–510. https://doi.org/10.1093/publius/pjae019

Yin, L. and Sankin, A. (2022, October 19). Dollars to Megabits, You May Be Paying 400 Times As Much As Your Neighbor for Internet Service. *The Markup*. https://the-markup.org/still-loading/2022/10/19/dollars-to-megabits-you-may-be-paying-400-times-as-much-as-your-neighbor-for-internet-service

APPENDIX TABLE 1.1 Recent Federal Funding Programs in Which Broadband Is an Eligible Use

Act	Program	Agency	Description
CARES (2020)	ReConnect Program	USDA	Allocates $100 million for grants, loans and grant-loan combinations for broadband deployment, upgrades and equipment in rural areas with populations of 20,000 or less. The program focuses on areas where 90% of the population lacks access to 10/1 Mbps Internet service. ReConnect-funded projects must meet the agency's speed requirements for new projects, 25/3 Mbps. Eligible applicants include public entities, territories and tribal governments, co-ops, and corporations.
	Coronavirus Relief Fund	Treasury	Allocates $150 billion to assist states, territories, local and tribal governments with necessary expenditures incurred during the pandemic. Broadband was an eligible use.
CAA (2021)	Broadband Infrastructure Program	NTIA	Allocates $300 million for broadband deployment, to public-private partnerships involving (1) a state or one or more political subdivision, and (2) a broadband provider. The program focuses on unserved areas, especially rural ones.

APPENDIX TABLE 1.1 (Continued)

Act	Program	Agency	Description
	Tribal Broadband Connectivity Grant Program	NTIA	Allocates $1 billion for tribal governments, tribal universities or colleges and tribal organizations, to support connectivity on tribal lands. Eligible uses include deployment, affordable broadband programs, remote learning, telehealth services, and digital inclusion efforts.
	Broadband Deployment Accuracy and Technology Availability Act	FCC	Allocates $65 million for the creation of broadband availability maps. The Act recognized the shortcomings of previous maps that relied on unverified, self-reported provider data that was collected at the census block level.
	Emergency Broadband Benefit	FCC	Allocates $3.2 billion to assist low-income families with a monthly $50-subsidy for monthly Internet service charges (up to $75 in tribal lands), and a one-time $100 payment for the cost of devices.
ARPA (2021)	Coronavirus State and Local Fiscal Recovery Funds	Treasury	Allocate $350 billion to assist states, territories, tribal and local governments with pandemic response efforts, including public health and economic recovery measures, the continuation of essential services, and long-term investments in public infrastructure, including broadband.
	Capital Projects Fund	Treasury	Allocates $10 billion to states, territories and tribal governments for critical capital projects that support in-person and remote work, learning, and health monitoring. CPF funds must be used for broadband projects where providers participated in the Affordable Connectivity Program, and that deliver equal to, or higher than 100/20 Mbps (Varn et al., 2023).
	Emergency Connectivity Fund	FCC	Provides an additional $7.171bn to schools and libraries for providing free access to broadband Internet and connected devices.

(*Continued*)

APPENDIX TABLE 1.1 (Continued)

Act	Program	Agency	Description
IIJA (2021)	Broadband Equity, Access and Deployment (BEAD)[1]	NTIA	Allocates $42.45bn to states, territories and tribal governments for broadband deployment grants. The program emphasizes unserved locations (access to less than 25/3 Mbps), underserved locations (access to less than 100/20 Mbps) and community anchor institutions without 1 Gbp Internet access.
			The amount allocated to each state was calculated based on their number of unserved locations and of "high-cost areas." The latter are census blocks where the average cost of deployment is higher than in other unserved census blocks. These areas are characterized by their remoteness, challenging topography, low-population density, and high poverty. At least 80% of locations must be unserved (Internet for All, n.d.)
	Digital Equity Act Programs	NTIA	Allocates $2.75 billion for the establishment of three programs that will help states, territories and tribal governments support the development and implementation of digital equity plans. The programs prioritize barriers to broadband access and adoption among populations that are disproportionately impacted by the digital divide – including rural, high-poverty, minority, and elderly individuals. The program was cancelled in May 2025.
	Tribal Connectivity Technical Amendments	NTIA	Allocates $2 billion to the Tribal Broadband Connectivity Program established through the Consolidated Appropriations Act of 2021.
	Enabling Middle-Mile Infrastructure	NTIA	Allocates $1 billion for middle-mile infrastructure deployment. Eligible entities include states, local and tribal governments; private telecommunications companies, telephone and broadband co-ops; electric utilities and electric co-ops, and public utilities.

APPENDIX TABLE 1.1 (Continued)

Act	Program	Agency	Description
	Affordable Connectivity Program (ACP)	FCC	Replaced the Emergency Broadband Benefit Program established through the Consolidated Appropriations Act of 2021. Allocates $14.42bn to assist low-income households with a monthly $30-subsidy for monthly Internet service charges (up to $75 in tribal lands). The program ran out of funds in May of 2024, and was not extended.
	ReConnect Program	USDA	Allocates $1.926 million for grants, loans and grant-loan combinations for broadband deployment, upgrades, and equipment in rural areas where 90% of the population does not have access to 100/20 Mbps. Eligible applicants include public entities, territories and tribal governments, co-ops, and corporations.
	Rural Broadband Program	USDA	Allocates $74 million for grants for broadband deployment and upgrades in rural areas where 50% of the population does not have access to at least 25/3 Mbps. Eligible applicants include public entities, territories and tribal governments, co-ops, and corporations.

Sources: Congressional Research Service (2021a), Congressional Research Service (2021b), Congressional Research Service (2021c), Council of State Governments (n.d.), Read and Wert (2021), The Pew Charitable Trusts (2020).

1 BEAD was restructured in June 2025 by the Trump Administration. See https://www.ntia.gov/other-publication/2025/bead-restructuring-policy-notice

2

FEDERAL-STATE REALIGNMENT OF BROADBAND AND DIGITAL EQUITY POLICY IN THE UNITED STATES

Johannes M. Bauer

Introduction

Broadband refers to Internet access of a quality that is sufficient to support advanced applications and services. Given that digital technology and its uses are continuously evolving, there is no stable threshold beyond which a network connection can be considered broadband. In an early effort to provide guidance, the Computer Science and Telecommunications Board offered two perspectives on how broadband could be operationalized. Starting from prevailing technologies and uses, it could be defined as the set of network capabilities that support the use of advanced applications. However, in a forward-looking perspective, broadband could also be conceptualized as the set of network capabilities needed to enable the development of next generations of applications and services (CSTB, 2002).

Network infrastructure deployment and upgrades require high investment and take time. Thus, a country seeking to benefit from the expanding opportunities offered by advanced information and communication technologies must select an upgrade path that balances the availability of the most advanced network technologies and the capabilities needed by all. Given that network infrastructure has strong public good characteristics, it is likely that simple reliance on decentralized market decisions will not result in the desirable balance. A well-designed legal and regulatory framework is an important contribution to supporting decisions by private and public investors to align infrastructure deployment, adoption, and usage with connectivity goals. It can facilitate the deployment of a network that supports capabilities that are widely needed while providing sufficient incentives for the development of advanced features required to push the frontiers of innovation.

DOI: 10.4324/9781003619208-3

The most advanced connectivity will initially only be available to a smaller set of locations and users before it is deployed ubiquitously across the network infrastructure. Because of dynamically improving capabilities and the time and cost required to deploy upgraded capabilities, it is likely that some degree of digital inequality will occur. The cost of upgrades varies considerably by location, resulting in an uneven upgrade of network capabilities. Ubiquitous deployment and measures to support digital inclusion can help reduce social and economic inequality; uneven deployment and longer delays until upgrades are made will likely amplify social and economic inequality. How fast new technologies should be available to all Americans and which measures are needed to achieve an acceptable outcome over time are highly contested issues of U.S. broadband policy.

The transmission capacity of services that U.S. policymakers decided should be widely available increased from 200 kilobits per second (Kbps) upload and/ or download speed in 1997 to 100 megabits per second (Mbps) download speed and 20 Mbps upload speed in 2024 (FCC, 2024a; Kruger, 2017). These adjustments reflect changes in applications, services, devices, and use patterns, but there is no systematic way to establish these thresholds, and the changes were somewhat haphazard and pragmatic. Multiple fixed and wireless technologies are available that can deliver these speeds in broadband access networks. These include coaxial cable, a technology utilized by cable television companies since the 1990s to provide broadband access, fiber optical networks, terrestrial and satellite-based licensed wireless services, and unlicensed wireless technologies (e.g., Wi-Fi, WiMAX) (FCC, 2022; Oughton et al., 2024).

Depending on the use case, these access technologies may compete (e.g., cable and fiber Internet access can support remote work) or complement each other (e.g., users who can afford it typically subscribe to fixed and mobile broadband). The technologies have different capabilities and cost characteristics and are subject to divergent regulation. For example, fiber technology supports exceedingly high transmission capacity and is therefore scalable to adjust to future demand growth. It is expensive to deploy, but once installed, it requires only modest maintenance and operations cost. In contrast, terrestrial wireless access technologies support somewhat lower transmission capacities, require lower investment, but have higher operations and maintenance costs. Consequently, the advanced communications infrastructure is an assemblage of heterogeneous technologies that evolves gradually and often in an unbalanced way.

Already in the 1930s, in the preamble of the Communications Act of 1934, the United States established a powerful overarching vision

> … to make available, so far as possible, to all the people of the United States, without discrimination on the basis of race, color, religion, national

origin, or sex, a rapid, efficient, Nation-wide, and world-wide wire and radio communication service with adequate facilities at reasonable charges … (Pub. L. 73–416, Section 1).

These principles were implemented in an increasingly complicated system of transfers between markets and customer segments. When the Telecommunications Act of 1996 introduced competition to local voice markets, the system required an overhaul. It was not until the early 2000s that support was gradually extended to broadband access.

Two crises, the Great Recession of 2008 and the COVID-19 pandemic, heightened awareness of the digital divides in broadband access that had been pointed out by the National Telecommunications and Information Administration since 1995 (NTIA, 1995; NTIA, 1998; NTIA, 1999; NTIA, 2000). They resulted in new federal and state policy initiatives that were specifically targeted to broadband connectivity. In the Infrastructure Investment and Jobs Act of 2021 (IIJA), U.S. Congress articulated digital equity as an overarching vision and goal for broadband policy and appropriated $65B to support broadband deployment and adoption. For important programs, particularly the Broadband Equity, Access and Deployment Program (BEAD), the U.S. Congress adopted a policy model that delegated important implementation issues to the states. With a new administration arriving in 2025, changes to the existing programs can be expected.

The next section of this chapter discusses the unraveling of the historical policy regime put in place to guide universal service policy. Next, I analyze the recent proliferation of broadband initiatives and the emerging new division of labor between federal, state, and local players. State and local governments have gradually taken on a stronger role during the past decade, and examples of this are provided in the rest of this book. The Bipartisan Infrastructure Bill further solidified that shift, and the following section discusses the greater emphasis given to digital equity and the contradictory technological, economic, and social forces that narrow and widen it.

The multiplication of goals, of programs, and of stakeholders raises new challenges for policy design and evaluation. I develop a conceptual framework of broadband policy as governing a dynamic, co-evolving socio-technical system. This allows positioning of the roles and intervention points of public policy and offers an analytical approach to assess the effectiveness of the plethora of co-existing governance measures. This chapter concludes with a brief discussion of potential ways forward, as the realignment of broadband policy will likely be affected by the change to a Republican administration.

The Weakening of the Historical Universal Service Regime

Effective, rational governance of broadband infrastructure requires the alignment of three interrelated factors: It must be based on a proper understanding of the system to be governed, it must have viable instruments available to attain the envisioned goals, and it must be politically feasible. Many current policy debates are related to finding a workable agreement on these dimensions and the difficulties of aligning them. As with other information and communication technologies before, structural and social changes during the past decades have undermined the effectiveness of the prevailing broadband policy practices. Realignment requires a fresh look, an adjustment of the existing tools, and the development of a new overarching approach (Bauer, 2022). It will also require renewed efforts to coordinate policies across levels of government and between the numerous stakeholders and public interest groups that have become involved in broadband policy.

Since the 1930s, the United States has relied on a decentralized system of telecommunications policy in which state and federal regulatory agencies contributed to assuring universal access to communications service. In the deregulatory spirit of the 1990s and with the Internet considered a borderless technology, federal policy seized the momentum and adopted a market-driven approach. The strengths and limitations of this policy approach have become visible during the past decade. In response, state and local governments have reclaimed policy initiative and adopted measures to narrow gaps in high-speed Internet access. Federal spending programs during the pandemic and the IIJA boosted these efforts by channeling significant funding to state and local programs.

Many of the instruments available to influence broadband policy evolved from earlier periods. Historically, the telephone system was governed by the common carrier principles encoded in Title II of the Communications Act of 1934 as amended. Like the cost of providing modern broadband infrastructure, the investment required to roll out telephone service varied widely between densely populated urban areas and sparsely populated rural areas, which were typically much more costly to serve. In response, regulators and service providers over time developed a complicated system of internal cross-subsidies to support network expansion. Above-cost prices in urban areas and for certain types of service (e.g., long-distance voice calls, business services) generated surplus funds that could be used to subsidize rural areas and low-income subscribers.

When competition was introduced into long-distance and local services and numerous new players entered the market that were not vertically and geographically integrated, this system of internal cross-subsidies became unsustainable. Policymakers responded by introducing a new system of intercarrier compensation (FCC, 2011; Rosenberg et al., 2006) and a new funding model for universal service (Gilroy, 2011). Authorized by the Telecommunications Act of 1996, the FCC in 1997 created a Universal Service Fund and the Universal Service Administrative Company (USAC), an independent non-profit company, to implement the new universal service policies, primarily geared to voice services. Four main programs were created to fund high-cost areas, low-income populations (Lifeline), and special programs to improve access to advanced telecommunications for schools and libraries (E-Rate), and rural hospitals. Several states developed their own universal service programs to support access to telephone service.

The first generation of dial-up Internet services operated over the telephone network. Thus, voice universal service policy indirectly also facilitated wider adoption of Internet access. Internet Service Providers (ISPs) benefited greatly from the wide adoption of telephone services and the non-discrimination provisions embedded in the common carrier model. Together with the newly liberalized market for telecommunications equipment they allowed decentralized user groups and online service providers to freely configure online services on top of the telephone network (Driscoll, 2022). Online services and first-generation dial-up Internet services expanded swiftly. As the connection speed supported by modems and voice networks gradually increased, new and innovative services could be offered. Network infrastructure and services evolved in a mutually enforcing, synergistic fashion. Higher speed networks could have evolved in the same governance model. However, policy changes in the 1990s put broadband on a different course.

Drafted during the height of belief in the benefits of unregulated market forces, the Telecommunications Act of 1996, a major overhaul of communications legislation, introduced competition as the overarching organizational principle for the communications sectors. In that spirit, it established that the Internet should remain "unfettered from state and federal regulation." In the late 1990s, entrepreneurial cable television companies had started to digitize their networks and offer higher speed Internet access to diversify their entertainment revenue streams. This created a bifurcation in the regulation of Internet access services. If provided by telephone companies, Internet access was treated as a common carrier service (codified in Title II of the Communications Act), subject to numerous restrictions and obligations. In contrast, if provided by cable companies, it was treated as an information service operated under a much more flexible, light-handed regulatory framework (codified in Title I of the Communications Act).

The FCC could have reconciled these discrepancies by classifying cable Internet service as a Title II common carrier service. Instead, it affirmed its position that cable modem services should be treated as an essentially unregulated information service according to Title I of the Act. In the *Brand X* case this issue made it all the way to the U.S. Supreme Court. Decided in 2005, the Court did not rule on the merits of the case but rather affirmed the power of the FCC to classify communications services as either common carrier (Title II) or information service (Title I). In its decision, the Court relied on the then intact Chevron defense, which implied deference to an expert agency in matters where the law was ambiguous.[1] With a Republican majority, the FCC subsequently reclassified other broadband access technologies (digital subscriber line (DSL), wireless broadband, and broadband over powerline (BPL)) as information services.

As part of these reforms, policy action shifted from the state to the federal level. Title II, common carrier regulation, had evolved over decades under joint federal and state regulatory oversight. Following the constitutional model of the United States, the FCC was responsible for interstate and international issues and state regulatory commissions for intra-state matters. The telephone network and cable networks fit into this spatial model of regulatory cooperation. However, the Internet, a logical network of networks that integrates a patchwork of heterogeneous physical networks into a seamless, borderless communications platform did not fit this model. Moreover, the policy vision embraced in the 1990s was to keep the Internet largely free of government oversight. This regulatory philosophy and the reclassification of broadband access as information services greatly reduced the role of traditional state regulation in broadband.

Federal universal funding programs were historically designed for common carrier voice services. They contributed to high-speed Internet access development only indirectly and haphazardly. For example, subsidies to carriers installing phone lines in high-cost areas could also be used to provide DSL broadband access. Aware of the increasing importance of advanced communications, Section 706 of the Telecommunications Act of 1996 had authorized the FCC and State commissions with jurisdiction over telecommunications to

> encourage the deployment on a reasonable and timely basis of advanced telecommunications capability to all Americans (including, in particular, elementary and secondary schools and classrooms) by utilizing, in a manner consistent with the public interest, convenience, and necessity, price cap regulation, regulatory forbearance, measures that promote competition in the local telecommunications market, or other regulating methods that remove barriers to infrastructure investment (47 U.S. Code § 1302).

Advanced telecommunications capability was defined as "high-speed, switched, broadband telecommunications capability that enables users to originate and receive high quality voice, data, graphics, and video telecommunications using any technology." The Act required the FCC to conduct annual inquiries into the availability of broadband and whether it was deployed in a reasonable and timely fashion (referred to as Section 706 Reports, e.g., FCC, 2024a; Kruger, 2017). The Act also established a framework to assess whether universal service funding mechanisms should be adapted to support advanced telecommunications in addition to voice services. Section 254 of the Telecommunications Act instructed the FCC to establish a Federal-State Joint Board on Universal Service (hereinafter "Board").

In its initial recommendations in 1997, the Board did not recommend extending the existing universal service mechanisms, particularly Lifeline, to retail broadband access because it was not yet widely adopted by the population.[2]

Hopes were high that market-driven, private sector investment would result in rapid deployment of advanced telecommunications capabilities. Considering that the emerging Internet economy boosted innovation, it was difficult to anticipate which upgrades in transmission capacity and the quality of broadband services would be needed to sustain innovation. The diversification of largely unregulated cable companies into broadband provision and the reclassification of other broadband Internet access services as lightly regulated information services were expected to further accelerate network deployment.

Moreover, in an increasingly polarized political landscape, broadband policy took a stronger ideological turn. The 1997 position of the Joint Board was a possible interpretation of the empirical evidence at the time as Internet access was primarily provided over the universally available telephone network. In hindsight, however, the assessment failed to anticipate the trajectory of broadband and the limitations of market forces to provide universal broadband service. Between 1999 and 2008, under Republican leadership, the FCC concluded in the congressionally mandated Section 706 Reports that broadband was deployed to all Americans in a reasonable and timely fashion (Kruger, 2017, pp. 8–9). This conclusion was based on the original definition of broadband as a transmission capacity of 200 Kbps at least in one direction (upload and/or download).

A Proliferation of Broadband Initiatives

While the FCC took a hands-off approach, state and local, public and private organizations recognized the potential benefits of appropriate broadband infrastructure initiatives, and some assumed a more initiative-taking role. Foundations, such as the Blandin Foundation in Minnesota, and non-profit organizations, such as the California Emerging Technology Fund (CETF), slowly started to work with community leaders to strengthen broadband access and digital inclusion.[3] Non-profit organizations such as the Alliance for Public Technology (APT, active 1989–2009) and the Benton Foundation (established in 1981, since 2019 renamed to Benton Institute for Broadband and Society) started to advocate for universal broadband access. And public-private collaborations, such as Connected Nation and its state affiliates, an initiative dating back to 2001 and primarily supported by private industry and other stakeholders, started to work with state and local planners to advance broadband.

The financial crisis of 2008 and the pandemic in 2020, jolted federal and state programs into a more active mode. In the wake of the Great Recession of 2008, U.S. Congress passed the American Reinvestment and Recovery Act of 2009 (ARRA), which directed the FCC to expand universal service programs to broadband. It also included significant federal funding for the expansion of broadband access. ARRA appropriated $4.7 billion for the National Telecommunications and Information Administration (NTIA) to increase broadband access and adoption. In addition, it appropriated $2.5 billion for the Rural Utility Service (RUS) in the U.S. Department of Agriculture. The FCC was entrusted to develop a comprehensive broadband plan, which was released in 2010 after an inclusive stakeholder consultation process (FCC, 2010). Finally, the Act appropriated funds to put together a national broadband map to guide programs.

With this broader mandate and in response to increasing use of video online, the FCC began to define the quality threshold that constituted a broadband connection more aggressively. In 2010, broadband was redefined as a connection supporting at least 10/1 Mbps and in 2015 it was again increased to 25/3, the current official standard. Forward-looking programs such as BEAD aim at 100/20 and 100/100 Mbps. Whenever the threshold is increased, the extent of digital inequality is affected also. Consequently, from 2010 onward the FCC periodic reviews of broadband deployment concluded that it was not reasonable or timely. This continuous adaptation created the challenge that legacy programs, which are typically designed with a five-to-ten-year timeline, may continue to support connectivity that is below the new threshold.

New programs built on the existing funding mechanisms for voice services and the regulatory rules governing intercarrier compensation (Kruger & Gilroy, 2019; Rosenberg et al., 2006). However, in the new competition-driven environment, the FCC gradually moved away from cost-based approaches to programs that embraced market mechanisms to achieve higher efficiency. Interventions were targeted narrowly to specific areas, functioning as stopgaps in areas which competition and private entrepreneurial initiative did not reach. In 2011, a new Connect America Fund (CAF) replaced the traditional High-Cost Fund and established several modernized programs to support the expansion of broadband with innovative instruments (FCC, 2011). These include model-based support (i.e., subsidies based on a national benchmark rather than actual cost) and reverse auctions (i.e., the winning bid goes to the operator with the lowest subsidy need).

At the same time, the number of parallel funding programs exploded from four to 17, often with widely differing eligibility criteria. High-cost support programs and later the programs in the Connect America Fund typically operate over ten-year windows. It is therefore possible, even likely, that older programs that have lower speed thresholds continue to be funded in parallel to new programs. For example, between 2016 and 2020, the consumer-side Lifeline program supported broadband speeds that were below the 25/3 mbps threshold that had been adopted in 2015. From December 1, 2016, to December 1, 2017, 10/1 Mbps was accepted and from December 1, 2017, to December 1, 2018, 15/2 Mbps. In certain cases, subscribers qualified if they purchased 4/1 Mbps service. From December 1, 2018, to December 1, 2019, 18/2 Mbps; and from December 1, 2019, to December 1, 2020, 20/3 Mbps. Beginning December 1, 2020, the then-prevailing 25/3 Mbps threshold was applied (see https://www.fcc.gov/general/lifeline-program-low-income-consumers.).

In August of 2019, the FCC replaced CAF with a new $20.4 B Rural Digital Opportunity Fund (RDOF) "to bring high-speed fixed broadband service to rural homes and small businesses that lack it" (see https://www.fcc.gov/auction/904). The reverse auction for the first tranche of support took place from October through November 2020. The effectiveness of CAF and RDOF programs was undermined by known flaws in the 2009 version of the national broadband map. Because of simplifying statistical assumptions, it overestimated broadband availability in rural areas and misrepresented broadband adoption in low-income urban neighborhoods. This contributed to a misallocation of subsidies. In 2020, in the Broadband Deployment Accuracy and Technological Availability Act (Broadband DATA Act), the FCC received funding to correct the errors. A new, more accurate map ("Broadband Fabric"), based on single serviceable locations, was released at the end of 2022 and is regularly updated.

Traditional programs to advance broadband were overshadowed by the COVID-19 pandemic and the demand for rapid action it generated. In rapid succession, the 2020 Coronavirus Aid, Relief, and Economic Security Act (Pub. L. No 116–136, CARES Act), the 2021 Consolidated Appropriations Act (Pub. L. No: 116–260), and the American Rescue Plan Act of 2021 (Pub. L. No. 117–2) earmarked more than $10B for broadband-related programs, much of it for services (e.g., tele-health, tele-education), and to support low-income households that could not afford broadband with the Emergency Broadband Benefit (EBB) program. The biggest boost for broadband was generated by the IIJA, which appropriated $65B for supply and demand-side broadband programs.

One program, the Affordable Connectivity Program (ACP), funded with $14.4B, was entrusted to the FCC. Although widely considered successful, there was also widespread agreement that ACP could have been designed more efficiently (Galperin & Bar, 2024). By February 2024, ACP supported 22.9 million non-tribal and 330,000 Tribal households.[4] However, the program drew down the available funding by June 2024 and U.S. Congress could not agree on a new appropriation (Horrigan, 2024).

The largest program established by IIJA, the Broadband Equity, Access, and Deployment Program (BEAD), appropriated $42.45 billion to NTIA for subsidies and grants to the states. The Digital Equity Act (DEA) designated $2.75B for three programs that provide funding to promote digital inclusion and advance equity for all. An additional $2B was designated to support connectivity on tribal lands and $1B to upgrade enabling middle mile broadband infrastructure.

The IIJA introduced an innovative approach to intergovernmental collaboration that might also help overcome problems of decentralized knowledge. BEAD, administered by NTIA, allocated block grants to states in June 2023 that were prorated according to the share of unserved locations in a particular state in all unserved locations nationwide.[5] By December 2024, all 56 states and territories had their initial proposals for how to implement BEAD at the state level approved by NTIA, 15 states had begun the process of selecting sub-awardees, and three states (Louisiana, Delaware, and Nevada) had completed the selection of sub-awardees.[6]

However, IIJA and the subsequent Notices of Funding Opportunity (NOFOs) did not completely decentralize responsibility. All programs must be developed under close supervision by federal agencies, something that many states are not used to. Each program has a planning phase, a review, commenting, and approval phase during which NTIA can require changes to the proposed plan, and an implementation and review phase. The grant programs establish requirements to monitor "measurable outcomes." For example, IIJA requires that states monitor five outcomes, including the availability of, and affordability of access to, fixed and wireless broadband technology, the online accessibility and inclusivity of public resources and services, and digital literacy. The Act also requires periodic evaluations of the broader community outcomes of improved digital connectivity (Knittel et al., 2024).

In parallel to these federal initiatives, states, municipalities, cooperatives, advocacy groups, and citizen initiatives, which were concerned about the uneven pace of broadband expansion, started to respond to local needs and the shortcomings of federal policy with initiatives to narrow connectivity gaps in a more timely manner (Ali, 2021; Strover et al., 2021). Some of these initiatives utilized legacy agencies, such as state regulatory bodies, but many resulted in the creation of new organizations and programs. Within a decade, broadband policy again developed into a multi-centric, nested system of local, state, and federal players, with considerable bottom-up momentum. By 2022, all states had some form of broadband program. We examine 17 state programs from 2014 to 2022 in Chapter 3 of this book.

Municipalities and townships remained involved in managing rights of way in many states, which gave them some influence over telecommunications development. Recognizing the Internet as an increasingly critical infrastructure, states and local communities, often supported by private foundations, became involved again in shaping broadband policy. Although nearly 30% of U.S. states prohibit or impede public and municipal ownership,[7] electric cooperatives and municipalities started to fill gaps that were not closed by private enterprises. See more on this in the case study Chapters 4–8. By 2024, more than 600 municipalities and upwards of 300 cooperatives were offering Internet access services. Public and cooperative enterprises often pursue a broader set of objectives than private enterprises and thus may be better aligned with public interest goals.[8] Federal, state, and municipal initiatives were complemented by mandated and voluntary programs of private sector companies and foundations. For example, Comcast started to offer Internet Essentials, a low-cost Internet access service, in 2011, as a voluntary commitment made to facilitate obtaining FCC approval of the proposed merger with NBC Universal (Rosston & Wallsten, 2020).

Thus, by 2023, a new balance between federal, state, and local initiatives had emerged. Not only had many states and municipalities developed their own broadband initiatives, but the new federal programs also envisioned a partnership between levels of government that devolved the implementation of programs in the hope that this would allow better responses to varying local conditions, such as barriers to network deployment and user adoption. Local and state action had created a diverse landscape of approaches and broadband policy experiments, as described in the rest of this book. Current federal, state, and local broadband and digital equity policy initiatives seek to provide some overarching guidance to the plethora of local and state initiatives. It is not clear whether the coordinative push will be sufficient to achieve this or whether the pending programs will further diversify local approaches.

Challenges of Achieving and Safeguarding Digital Equity

In most current initiatives and debates, broadband access is seen as a necessary but not a sufficient condition for social inclusion and economic prosperity. Thus, the programs that emerged from the Bipartisan Infrastructure Bill recognize the importance of improving digital literacy, of providing means to support the maintenance and upgrade of devices used to access, process, and store information for realizing the potential benefits of broadband. As programmatic visions are translated into practical policy implementations, some of these nuances may be lost. Practitioners and funders typically work with simplified "logic models" that map how actions translate into outcomes. Often, a simple causation from broadband deployment to adoption, use, and beneficial community impacts is asserted (e.g., see the appendix in Rhinesmith et al., 2023). Whereas this is not necessarily wrong, it may overlook that digital equity is affected by factors that narrow digital inequalities and others that widen it. The versatility and dynamic development of digital technologies and their applications require continuous vigilance and monitoring of these contradictory forces and their net effect (see Figure 2.1).

Over time, it is possible that the emergence of new access technologies and devices may reopen digital divides. For example, rapid technological change that requires frequent device upgrades may create new access divides. The introduction of new digital services, such as telehealth or mobile banking, may exclude certain user populations and increase inequalities. Or the introduction of artificial intelligence applications that require high-quality computing support to run well may exclude many existing users who do not have the appropriate equipment and/or the resources to upgrade. A challenge in the present context of rapid technological change is, therefore, to shape technological developments in a human-centric direction. For beneficial services, it will be important to accelerate the adoption of services that are considered part of core infrastructure.

Factors narrowing digital divides and inequalities	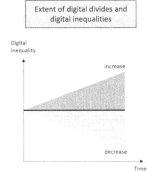	Factors widening digital divides and inequalities
• Cheaper services, devices • Policies, such as subsidies for network upgrades • Programs supporting affordability (e.g., ACP) • Digital literacy training in K-12, post-secondary, adult education • Upskilling and retraining of the workforce • Human-centric design of technology and services • Appropriate governance •		• More expensive services, devices • Policies, such as market-driven infrastructure roll-out • Migration of services to online delivery (e.g., during the pandemic) • Next-generation technologies (e.g., AI, IoT, robotics, metaverse, ...) • User-unaware technology and service design • Inappropriate governance • .....

FIGURE 2.1   Forces narrowing and widening digital inequalities.

Source: Author. Note: The level of digital inequalities is an outcome of forces working toward narrowing discrepancies and factors increasing them. For example, cheaper service and devices will narrow broadband access divides, all other things being equal. In contrast, higher access prices for service and costs of devices will have the opposite effect. At any point in time, the level of digital equality is an emergent outcome of these multiple, opposing forces in the socio-technical system.

Moreover, policy is challenged to provide training and education to develop the skills necessary to participate in technology-enabled activities.

Recent data published by the FCC (2022; 2024a) clearly show the challenges associated with the need to continuously upgrade network infrastructure. Inequality of access is, not surprisingly, more serious for higher access speeds. A key challenge, then, is for policy to find legal and institutional arrangements that allow upgrading the broadband access infrastructure continuously to these higher standards. The market-driven policy regime that was established with the Telecommunications Act of 1996 and subsequent regulatory policies accomplished this goal partially, but it also failed in predictable ways. Between 2000 and 2020, private sector companies invested more than $300 billion in fixed and wireless infrastructure upgrades. Urban areas, highly educated, high-income households benefited from the availability of world-class services.

The past three decades demonstrate that private entrepreneurship, although a potent force, did not suffice to expand service to high-cost, rural, and remote areas. Moreover, the incentives to provide service to marginalized urban and rural populations, even where networks were available, were insufficient. It is too early to assess whether the portfolio of new approaches will suffice to close current broadband access and skills gaps. In fact, there are reasonable concerns that currently there are too many programs, often with contradictory goals and incentives (GAO, 2015; GAO, 2022; GAO, 2023).

It is also too early to assess the cost-effectiveness of the programs. The massive amount of funding clearly will have an impact on rural connectivity. It will bring unserved areas online and improve the quality of service to underserved locations. Early estimates suggest it will suffice to connect all unserved locations in most states, except for highly rural states. Complementary programs to advance digital literacy, to provide workforce training, and to develop technical support programs should also move broadband connectivity in the right direction. See more discussion on this in Chapters 8–10.

Future Challenges

Although the Bipartisan Infrastructure Bill was supported by both parties in Congress, there has been mounting criticism on its implementation, including concerns about the built-in preference for fiber which may create additional delays for many locations, the heavy administrative burden, and the slow process. Much of the criticism comes from partisan stakeholders, such as libertarian think tanks, the satellite industry, and members of the Republican party. However, independent of political affiliation, a critical analysis of the current state of broadband policy shows several weaknesses that a forward-looking, efficient approach will have to address. Multiple, interrelated issues must be addressed, and several administrations have failed or shied away from trying. Not only are the issues contested, but there are also numerous veto players that can block solutions.

First, the current system of universal service support requires an urgent update both on the spending and contribution side. Several of the existing programs, particularly in the high-cost areas, are obsolete but other components are missing. ACP demonstrated the power of demand-side interventions to overcome challenges of affordability in areas that have network infrastructure available. The existing Lifeline program is insufficient to achieve that goal. A focused, means-tested approach would go a long way to close the remaining adoption gaps. States are implementing policies for low- and middle-income families, so any federal policy would have to be appropriately coordinated with state approaches. Once BEAD is completed, the need to subsidize capital expenditures will change and many network operators may need support for operations and maintenance expenses. No such program is currently available.

Moreover, the system of contributions to the Universal Service Fund is broken. The contribution factor for the first quarter of 2025 is set at 36.3% (of the voice revenues on which the contributions are assessed).[9] A 2021 white paper commissioned by three associations (Incompas, NTCA–The Rural Broadband Association, and the Schools Health & Libraries Broadband Coalition) proposed expanding the funding base to broadband Internet access services, which would greatly reduce the contribution factor (Mattey, 2021). Concerned about the regressive effect of such a measure on low- and middle-income households, the FCC opted to forebear from such a broadening in the 2024 Safeguarding and Securing the Open Internet Order (FCC, 2024b).[10] Other proposals, discussed in

Congressional meetings and by some stakeholders, would broaden the contribution base further to include content providers, especially "big tech" platforms. These proposals raise several conceptual and implementation concerns, and their advantages and disadvantages would have to be analyzed relative to other funding models.

Second, over the past two decades efforts to address specific, narrowly defined policy problems, such as expanding the network to specific locations, have resulted in a proliferation of programs. The Government Accountability Office (GAO) has identified more than 133 federal programs administered by 15 agencies (GAO, 2023). Of these, 25 programs have broadband as their main purpose. Targeting deployment, affordability, devices, digital skills, and planning, 13 of these programs overlap but their interaction is not well understood. Whereas it is possible that they complement each other, focusing on different problems, they could also duplicate efforts or compete. For example, there are concerns that some of the RDOF defaults may be the result of newer subsidy programs, such as BEAD, offering more attractive subsidies. Programs and eligibility rules were developed over multiple years and did not anticipate, and hence did not adopt any safeguards against such strategic behavior.

Third, it would be desirable to develop a more flexible approach to establish thresholds for policy. Rather than pick a throughput and quality standards somewhat haphazardly, zones characterized by different quality connectivity could be established. Upgrade plans could then devise differentiated measures and goals for these zones, without privileging certain technologies. This would allow returning to relatively technologically neutral broadband policy. It would allow reliance on bridge technologies that provide better than existing connectivity, even if they may not be long-term scalable solutions. Such a model would require new and innovative subsidy policy designs, but it would allow novel entrepreneurial solutions and most likely would shorten the time to better connectivity for many of the currently unserved locations. In December 2024, the Maine Connectivity Authority (MCA) opened enrollment to participate in the Working Internet ASAP (WIA) program designed to fund access to satellite Internet to 9,000 locations without any Internet access.[11] The program is envisioned as an intermediary step to provide quick connectivity while additional infrastructures are deployed. Such initiatives will need careful planning to minimize the total program cost over the planning horizon, but they would help overcome the predicament of individuals in unserved locations who might face years of wait times despite programs such as BEAD.

Democrats and Republicans recognize broadband as a critical infrastructure, but they disagree on the importance of achieving digital equity. They have divergent views as to which specific measures will best be suited to close the remaining connectivity gaps. These differences could lead to significant program changes at the federal level or even policy gridlock. At the state and local level, it will lead to variations in policy design within the overarching guidance from the federal government. Rigorously evaluated, this diversity could be translated into a dynamic learning system that could help inform forward-looking policies. Many current programs are time-limited and will expire in six to ten years. Federal and state policymakers will have a few years to devise a sustainable approach that assures that all Americans have access to connectivity of a quality that supports advanced applications and continuous upgrades of these capabilities to enable next-generation uses and applications. If they fail, the future will likely see a continuation of cycles of decreasing and increasing digital inequalities.

Notes

1 The U.S. Supreme Court overturned the Chevron doctrine in *Loper Bright Enterprises v. Raimondo* on June 28, 2024. Under Chevron, agencies were given deference if their policies could be considered "reasonable" interpretations of the law. Under the new standard, policies must be the "best" interpretation, a requirement that is more difficult to meet and that likely will reduce discretionary power of regulatory agencies.
2 Support generated by the other three programs also benefited broadband connectivity. High-cost funding for voice services also facilitated DSL Internet access, and E-Rate supported advanced telecommunications services to schools and libraries and their users, and the rural health care program connected hospitals and health care providers.
3 The Blandin Foundation (https://blandinfoundation.org/programs/broadband/) started to engage in broadband issues in 2003. The California Emerging Technology Fund was established in 2005 by the California Public Utilities Commission (CPUC), see https://www.cetfund.org/about-us/mission-and-history/.
4 See ACP Enrollment and Claims Tracker, retrieved on December 20, 2024, from https://www.usac.org/about/affordable-connectivity-program/acp-enrollment-and-claims-tracker/#total-households-at-enrollment-freeze.
5 See Biden-Harris Administration Announces State Allocations for $42.45 Billion High-Speed Internet Grant Program as Part of Investing in America Agenda, June 23, 2023. Retrieved on December 20, 2024, from https://www.ntia.gov/press-release/2023/biden-harris-administration-announces-state-allocations-4245-billion-high-speed-internet-grant.
6 See BEAD Progress Dashboard, https://www.internetforall.gov/bead-progress-dashboard. Retrieved on December 20, 2024, from https://www.internetforall.gov/bead-progress-dashboard.
7 See Sean Gonsalves, The State of State Preemption: Stalled – But Moving In More Competitive Direction, Community Networks, November 1, 2024. Retrieved on December 19, 2024, from https://communitynets.org/content/state-state-preemption-stalled-moving-more-competitive-direction.
8 At the same time, if they are not effectively managed, the public bears the risk of poor performance or failure. Moreover, there are concerns about safeguarding level playing fields between public and private players.
9 See Contribution Factor & Quarterly Filings – Universal Service Fund (USF) Management Support, retrieved on December 20, 2024, from https://www.fcc.gov/general/contribution-factor-quarterly-filings-universal-service-fund-usf-management-support.

10 The Order was challenged in the courts and on August 1, 2024, the U.S. Court of Appeals for the Sixth Circuit stayed the rules. A panel of three judges heard oral arguments on October 31, 2024. At the time of writing the decision is pending.
11 See Working Internet ASAP (WIA), retrieved on December 20, 2024, from https://www.maineconnectivity.org/wia.

References

Ali, C. (2021). *Farm Fresh Broadband: The Politics of Rural Connectivity*. MIT Press.

Bauer, J. M. (2022). Toward new guardrails for the information society. *Telecommunications Policy*, *46*(5), 102350. https://doi.org/10.1016/j.telpol.2022.102350

CSTB. (2002). *Bringing Home the Bits*. National Research Council, Computer Science and Telecommunications Board.

Driscoll, K. (2022). *The Modem World: A Prehistory of Social Media*. Yale University Press.

FCC. (2010). Connecting America: The National Broadband Plan. Washington, DC: Federal Communications Commission.

FCC. (2011). Connect America Fund et al., WC Docket No. 10–90 et al. Report and order and further notice of proposed rulemaking, FCC 11–161 at para. 1404 (rel. Nov. 18, 2011) (USF/ICC Transformation Order), Retrieved November 17, 2024, from https://docs.fcc.gov/public/attachments/FCC-11-161A1.pdf.

FCC. (2022). 2022 Communications marketplace eport. GN Docket No. 22–203, Report. Retrieved May 9, 2024, from https://www.fcc.gov/document/2022-communications-marketplace-report.

FCC. (2024a). Inquiry concerning the deployment of advanced telecommunications capability to all Americans in a reasonable and timely fashion (2024 Section 706 Report). Report. WC Docket No. 22–270. Retrieved August 6, 2024, from https://docs.fcc.gov/public/attachments/FCC-24-27A1.pdf.

FCC. (2024b). Safeguarding and securing the open internet; Restoring internet freedom, WC Docket Nos. 23–320, 17–108. Declaratory Ruling, Report and Order, Order, and Order on Reconsideration. Retrieved on December 20, 2024 from https://www.govinfo.gov/content/pkg/FR-2024-05-22/pdf/2024-10674.pdf.

Galperin, H., & Bar, F. (2024). Evaluating the impact of the affordable connectivity program.

GAO. (2015). Broadband: Intended outcomes and effectiveness of efforts to address adoption barriers are unclear. Report to Congressional Requesters. GAO-15-473. Washington, DC: United States Government Accountability Office.

GAO. (2022). Broadband: National strategy needed to guide federal efforts to reduce digital divide. GAO-22–104611. Washington, DC: U.S. Government Accountability Office.

GAO. (2023). Broadband: A national strategy needed to coordinate fragmented, over-lapping federal programs. GAO-23–106818. Washington, DC: U.S. Government Accountability Office.

Gilroy, A. A. (2011). Universal Service Fund: Background and options for reform. Congressional Research Service, Report RL33979. Washington, DC.

Horrigan, J. B. (2024). Leaving money on the table: The ACP's expiration means billions in lost savings: Benton Institute for Broadband & Society.

Knittel, M., Mack, E., Nam, A., & Bauer, J. M. (2024). Assessing the effects of the infrastructure investment and jobs act on high-speed internet access, digital equity, and community development. Opportunities and challenges of measuring the impact of broadband policy. East Lansing, MI: Quello Center at Michigan State University.

Kruger, L. G. (2017). Defining broadband: Minimum threshold speeds and broadband policy. Congressional Research Service Report R45039, https://sgp.fas.org/crs/misc/R45039.pdf.

Kruger, L. G., & Gilroy, A. A. (2019). Broadband Internet access and the digital divide: Federal assistance programs. Congressional Research Service Report RL30719 (Updated January 9, 2019). Washington, DC.

Mattey, C. (2021). USForward. FCC must reform USF contributions now: An analysis of the options. Bethesda, MD: Mattey Consulting in conjunction with Incompass, NTCA, and SHLB.

NTIA. (1995). Falling through the Net: A survey of "have nots" in rural and urban America. Washingtion, DC: U.S. Department of Commerce, National Telecommunications and Information Administration.

NTIA. (1998). Falling through the Net II: New data on the digital divide. Washington, DC: U.S. Department of Commerce, National Telecommunications and Information Administration.

NTIA. (1999). Falling through the Net III: Defining the digital divide (Vol. U.S. Department of Commerce, National Telecommunications and Information Administration): Washington, DC.

NTIA. (2000). Falling through the Net IV: Toward Digital Inclusion. Washington, DC: U.S. Department of Commerce, National Telecommunications and Information Administration.

Oughton, E., Geraci, G., Polese, M., Shah, V., Bubley, D., & Blue, S. (2024). Reviewing wireless broadband technologies in the peak smartphone era: 6G versus Wi-Fi 7 and 8. *Telecommunications Policy*, 102766. https://doi.org/https://doi.org/10.1016/j.telpol.2024.102766

Rhinesmith, C., Dagg, P. R., Bauer, J. M., Byrum, G., & Schill, A. (2023). *Digital Opportunities Compass: Metrics to Monitor, Evaluate, and Guide Broadband and Digital Equity Policy*. Working Paper, Ann Arbor, MI: Merit Network, Inc. and East Lansing, MI: Quello Center. Retrieved from https://quello.msu.edu/wp-content/uploads/2023/02/Digital-Opportunites-Compass-Paper-20220223.pdf.

Rosenberg, E., Pérez-Chavolla, L., & Liu, J. (2006). *Intercarrier Compensation and the Missoula Plan*. Commissioner Briefing Paper. Columbus, OH: National Regulatory Research Institute (NRRI).

Rosston, G. L., & Wallsten, S. J. (2020). Increasing low-income broadband adoption through private incentives. *Telecommunications Policy*, 44(9), 102020. https://doi.org/https://doi.org/10.1016/j.telpol.2020.102020

Strover, S., Riedl, M. J., & Dickey, S. (2021). Scoping new policy frameworks for local and community broadband networks. *Telecommunications Policy*, 45(10), 102171. https://doi.org/https://doi.org/10.1016/j.telpol.2021.102171

3

THE IMPORTANCE OF STATE POLICY DESIGN

Natassia A. Bravo and Mildred E. Warner

Introduction

The role of states in broadband funding was greatly supported by the American Recovery and Reinvestment Act of 2009 (ARRA). The Act allocated $4.7 billion to the National Telecommunications and Information Administration (NTIA) to assist states with deployment, adoption efforts, data collection, and broadband planning (Congressional Research Service, 2011). Some states used these funds to establish their own broadband programs, which later began awarding deployment grants (Lichtenberg, 2017). States design their eligibility criteria to ensure that funds support specific technology and speed requirements, to encourage participation from public-private partnerships (PPPs), and to support both last-mile and middle-mile deployment.

This chapter focuses on the role of state broadband programs in closing the digital divide. We briefly describe how current federal policy expands definitions of equity and access. Then we explore the lessons from earlier state policy efforts on the impacts of grants to expand digital access. Experience from these state efforts can help inform the new federal programs established by the Infrastructure Investment and Jobs Act of 2021 – particularly the Broadband Equity, Access and Deployment Program (BEAD).

Changes in Federal and State Broadband Funding

Digital Equity and Infrastructure Funding

Today, it is widely acknowledged that the digital divide will not be closed by infrastructure rollout alone – and other barriers to adoption must be simultaneously

DOI: 10.4324/9781003619208-4

addressed. If a family is unable to afford a basic Internet subscription, it matters less if they have access to cable or fiber optic Internet service in their area. If a significant percentage of the population lacks knowledge on how to use the Internet to their benefit, there is little incentive for them to pay for a subscription. Unsurprisingly, federal policy began shifting away from a narrow focus on "access" to the broader goal of digital equity. The Digital Equity Act of 2021 acknowledged the driving role of socioeconomic inequality in preserving the adoption gap. The Act provided funding for broadband planning and adoption efforts, and requests that states focus on population groups that are disproportionately impacted by digital exclusion – such as rural, low-income, minority, and senior individuals (S.1167, 2019). In contrast, federal and state programs that fund broadband infrastructure tend to give greater focus to factors such as the percentage of unserved households and population density.

Why does socioeconomic inequality matter for deployment? Undoubtedly, there is little incentive for providers to build in rural and/or low-density areas, where profits are expected to be lower. These areas are already a priority for federal and state funding programs. Yet, as reports of digital discrimination suggest, there is little incentive for providers to build new or upgrade existing infrastructure in low-income (and minority) urban neighborhoods either. The business case is less attractive because the rate of subscriptions is expected to be lower, be it due to the cost of service or due to lack of digital literacy. Even with the incentive of funding, under-resourced and rural communities struggle to attract traditional providers, navigate grant applications, and provide local matches. This puts them at a disadvantage when competing for funding with projects in more profitable areas.

Population density is not the only determinant of access. Digital equity must also be considered in infrastructure policy, or else the divide is bound to persist. While the future of digital equity under the new Trump administration is uncertain, inequality remains a driver of the digital divide. States can fill this policy vacuum.

This chapter focuses on state broadband policy, and its role in closing the infrastructure gap. While universal access is the common goal across state broadband programs, each defines which types of providers, technologies, and areas are eligible and prioritized for grants. Could these programs also factor-in socioeconomic inequality in broadband infrastructure funding as well?

This chapter will use some of the demographic and socioeconomic factors linked to the digital divide – including population density, race, education, poverty, and rurality – to analyze how states distributed grants in the years before the pandemic. This will help us understand if states are ready to address the broader expectations of digital equity set out by the new funds from BEAD.[1] While states are expected to establish their own rules over the use of BEAD funds, they will need to comply with the NTIA's broader criteria and rules.

For the NTIA, BEAD presented an opportunity to address the shortcomings of past federal broadband funding programs, including minimum speed requirements, technology eligibility, and match requirements. First, the NTIA raised the speed requirements to 100/20 Mbps and prioritized fiber optic. Second, BEAD is open to "non-traditional providers" like PPPs, cooperatives, public utilities, non-profit organizations, and municipal broadband services. This encourages states, where local governments are prevented from owning and operating their own broadband networks, to broaden municipal authority (Ali et al., 2024). Third, BEAD requires a 25% match at minimum, but the program does allow applicants in areas where it is costly to build to request a waiver (NTIA, 2022).

Our analysis explores if states are ready to meet these changes, based on how 17 states awarded funds in the years before the pandemic.

What Can We Learn from State Broadband Programs?

The Pew Charitable Trusts collected data on 724 state broadband grants across 17 states on broadband grants awarded between 2014 and 2020. We looked at the characteristics of funded projects: provider and grantee type, technology type, metro/non-metro status, and whether these projects included middle-mile infrastructure and/or connected to a local planning process (Bravo and Warner, 2023; Bravo, 2025). Figure 3.1 shows the states included in our analysis, and the number of grants awarded per state.

We explore the types of projects that were funded. We know the digital divide intersects with urban-rural, socioeconomic, and racial inequities, but do these factors influence state grant distribution? States might be reluctant to include additional metrics of access and adoption in their criteria, lest they become barriers in the application process. By narrowly focusing on percentage of unserved and underserved, have states been able to reach rural, low-density, and high-poverty

FIGURE 3.1   State broadband programs analyzed. Image by authors.

Source: Pew Charitable Trusts State Broadband Grants, 2014–2020.

areas as well? Due to their limited fiscal capacity, these areas will not only struggle to attract for-profit providers, but also to meet grant requirements. Could state policies hinder access to broadband funding? What impact do state policy criteria have on grant distribution? How will states ensure funds reach areas disproportionately impacted by the digital divide?

We incorporate in our analysis several digital equity criteria that are mentioned in DEA and BEAD – population size, population density, race, education, poverty level, adoption, and provider availability. We expect state-level policies also to influence how funding is allocated, so we include the state program's required match and whether the state preempts municipal broadband.

How Do State Policies Differ?

The Pew Charitable Trusts collected data on the eligibility criteria, requirements, and priorities of state broadband programs. We added information on policy design from documents publicly available at these state program's websites. Differences in state policy are described in Table 3.1.

Ten states required a local match, which ranged from a low of 20% in Indiana to a high of 85% in Iowa. Most states were technology neutral, but four made satellite ineligible. Minimum upload/download speeds were 10/1, with some states requiring faster speeds. Seven states encouraged or required PPPs. Eleven states encouraged anchor institutions, such as hospitals and schools, be included; and eight encouraged a local planning process. In nine states, local governments which received federal funds were ineligible for state funds, while five states allowed localities to use federal funds to meet state match requirements. Ten of the 17 states restricted municipal broadband during this period. Washington lifted their municipal broadband restrictions in 2021, followed by Colorado (in 2023) and Minnesota (in 2024). We incorporate two of these state-level policies (required match percentage and municipal broadband restrictions) in our analysis of the counties which were awarded state funding.

How Did States Approach Broadband Funding?

Infrastructure Eligibility

States Are Technology Neutral, but Still Prioritize Fiber Optic Broadband

As we explained in Chapter 1, in the past, technology neutrality has been the norm for federal and state broadband programs. No broadband technology was to be prioritized over another as long as it could fulfill the program's minimum download/upload speeds and scalability. Thus, over the years, federal and state funds have been used to deploy a wide range of broadband technologies, including:

fiber optic, coaxial cable, Digital Subscriber Line (DSL), fixed wireless, satellite, or a combination of two technologies.

Not all broadband technologies are equal. Fiber optic broadband is an attractive long-term investment due to its capacity, reliability, and scalability (Afflerbach, 2022). However, it is not always a viable choice for rural and low-density areas, due to the low rate of return. In fact, the easiest, quickest, and less expensive approach might just be to subsidize older technologies for which there is already a large footprint, like DSL and cable. A larger number of users could be reached at a lower cost, which increases a project's chances to secure a grant. Predictably, early rounds of federal funding allocated billions of dollars to large telecommunications and cable providers that already owned much of the DSL and cable infrastructure in the country (Ali, 2021).

Unfortunately, older technologies are gradually becoming obsolete. They tend to be slower, less reliable, and less capable of sustaining increasing traffic than fiber optic broadband. When the FCC raised the minimum download/upload speeds required from 10/1 Mbps to 25/3 Mbps in 2015, DSL and older generations of satellite became obsolete and potentially ineligible for broadband funding. While other federal and state programs are not required to follow the FCC's footsteps, criticism is to be expected if later eligibility criteria result in billions of dollars wasted on obsolete infrastructure. Indeed, until recently, federal funds were used to provide service with speeds as low as 10/1 Mbps in rural areas (Neenan, 2023). If the rules are stricter, then funds are likelier to be used only for high-capacity infrastructure that will stand the test of time better than DSL, satellite, or cable.

Technology neutrality might be seen as a policy barrier to quality broadband service. Critics warn that it could lead to a waste of public dollars in substandard service (Ali, 2021; Dawson, 2023). The issue has drawn attention once again, as states prepare to award BEAD funds (Connected Nation, 2023). As Table 3.1 shows, the majority of state broadband programs in our study are technology neutral. Will states retain this policy, or will they openly prioritize fiber optic infrastructure? We look at how state broadband grants were allocated before the pandemic. Table 3.2 shows which technologies our 17 states funded between 2014 and 2020.

To our surprise, states were ahead of the curve and most grants supported fiber optic broadband projects. This result holds across urban, suburban, and rural areas. Fiber optic broadband projects also received more funds per premise passed ($3,366) than their non-fiber-optic counterparts ($1,921), which we attribute to the higher costs of deployment associated with fiber. Nonetheless, states made the effort to prioritize higher capacity, scalability, and reliability over cost. Less than 7% of grants were used to build DSL, satellite, or a combination of technologies. This is unsurprising, considering that half of states in our study had a minimum requirement of 25/3 Mbps and/or do not fund satellite (Table 3.1).

TABLE 3.1 State Broadband Program Policies, 2020, 17 States

State Program	Infrastructure Eligibility			Applicant/Provider Eligibility			Area Eligibility	Required Match (%)
	Program Is Open to All Technologies	Minimum Download/Upload Speeds (Mbps)	Middle-Mile Costs Can Be Included	Must Be a Public-Private Partnership	Electric Cooperatives Are Eligible	Municipal Networks Are Eligible	Program Requires that Project Area Meets One or more Criteria of Digital Equity (e.g., Rural, Low Population Density, High Poverty, etc.)	
AL	Yes	10/1	Yes	Not required	Yes	Yes	Forty percent of funds must be used in unincorporated areas.	65
CA	Yes	6/1	Yes	Not required	Yes	No	Service area must be rural. Additional funds go to areas where 80% of households are low-income.	0–40
CO-DORA	Not specified.	25/3	Yes	Not required	Not specified	Yes	Projects in low-density areas are prioritized.	25
CO-DOLA	Yes	Not specified	Required	Required			Program assists communities impacted by mineral and mineral fuel mining.	50
IN	Not satellite.	50/5	Yes.	Not required	Yes	No	Not specified	20
IA	Not specified.	25/3	Not specified	Not required	Yes	Yes	Projects serving low-density areas are prioritized.	85
ME	Yes	10/10	Yes	Prioritized.	Not specified.	No	Not specified	25

MA-LMIG	Not specified.	25/3	No	Not required	Not specified.	No	Not specified	N/A
MA-BEP	Only cable.	25/3	Not specified	Encouraged	Not specified	Yes	Program is for communities with less than 96% residential cable coverage.	N/A
MN	Not specified.	25/3	Yes	Encouraged	Yes	Yes	Not specified	50
MO	Yes	25/3	Yes	Encouraged	Yes	Yes	Areas with high poverty or unemployment are prioritized.	50
NC	Not satellite.	10/1	No	Not required	Yes	Yes	Projects in economically distressed counties are prioritized.	35–55
OR	N/A	N/A	N/A	N/A	Not specified.	No	N/A	N/A
SD	N/A	25/3	N/A	No	Not specified	No	N/A	50
TN	Yes	10/1	Yes	Not required	Yes	Yes	Projects in economically distressed counties are prioritized.	50
VT	Not satellite.	10/1	Not specified	No	Yes	No	Census blocks lack access to at least 4/1 Mbps are eligible.	N/A
VA-VATI	Yes	10/3	Yes	Required	Yes	Yes	Not specified	20
VA-TRRC	Yes	10/1	Yes	Required	Yes	Yes	Eligibility restricted to forty tobacco-dependent counties.	50
WA	Yes	50/10	Not specified	Required	Yes	Yes	Service area must be rural.	25
WI	N/A	25/3	Not specified	Encouraged	Yes	Yes	Not specified	0

Sources: (1) The Pew Charitable Trusts State Broadband Program Data (2020), (2) Temple University Center for Public Health Law Research's data on State Preemption Laws, (3) Publicly available information at these state program's websites. A list is included under References. Colorado, Massachusetts and Virginia each have two programs included in our study.

The policies featured in Table 3.1 were valid during the study period (2014–2020).

TABLE 3.2 State Broadband Grants by Technology and Metro Status

Technology Supported	Metro Status					
	Urban	Suburban	Rural	Unknown	Total	
	N	N	N	N	N	(%)
Fiber optic infrastructure	148	167	90	6	**411**	(57)
Unknown	35	47	32	0	**114**	(16)
Fixed wireless infrastructure	33	35	20	5	**93**	(13)
Cable infrastructure	31	11	10	0	**52**	(7)
DSL infrastructure	18	5	12	0	**35**	(5)
Combination of technologies	5	2	10	0	**17**	(2)
Another technology (e.g., satellite)	0	2	0	0	**2**	(0)
Total	**270 (37%)**	**269 (37%)**	**174 (24%)**	**11 (2%)**	**724**	(100)

Source: Author analysis of Pew Charitable Trusts State Broadband Grants 2014–2020, N = 724 funded projects in 17 states.

One might wonder: If states were already inclined to fund fiber optic, why not just break away from technology neutrality? For one, states might prefer to keep eligibility criteria broad, to encourage investment in less profitable areas. They might also be avoiding the current debate surrounding BEAD, which encourages states to prioritize fiber deployment whenever possible. BEAD's preference for fiber optic raised alarm among fixed wireless providers (Dano, 2022), but fixed wireless was the second most-funded technology by states before the pandemic, as Table 3.2 shows.

Fixed wireless technology has important limitations. To accommodate traffic and achieve high speeds, fixed wireless service relies on close proximity to cell towers and antennas, which must be densely deployed. Service can be disrupted due to signal interference from foliage, weather, and buildings. Equipment requires frequent upgrading as technology evolves (Afflerbach, 2022). Nonetheless, fixed wireless or satellite technologies are sometimes the only viable alternative for remote areas or challenging terrains (like mountainous regions or islands) where wired technologies like cable or fiber optic are too costly or physically impossible to deploy. We will see examples of this in Chapter 8 where we discuss broadband investment in remote Indigenous communities in Alaska.

It remains to be seen if fixed wireless and satellite will receive BEAD funding (Abarinova, 2024). Will states continue to fund these technologies, or shift away from technology neutrality? Regardless, future state eligibility criteria will need to match BEAD's definitions of "unserved" (without access to download/upload speeds of at least 25/3 Mbps) and "underserved" (without access to service of at least 100/20 Mbps). Some states have already been doing this. At least eight states already required projects to deliver speeds of at least 25/3 Mbps or higher during the 2014–2020 period. Since then, some states have raised their standards. For example, Alabama, Minnesota, and Indiana now prioritize projects that deliver service speeds of 100/20 Mbps or higher.

Middle-Mile Infrastructure Is Critical for Rural Connectivity, but only a Few Projects Include It

Broadband infrastructure has three layers: the network core, the middle mile, and the last mile. Federal and state programs have primarily focused on the deployment of last-mile networks, which provide retail service to homes and businesses. However, to connect to the broader Internet, last-mile providers must also link to a middle-mile network. Access to a high-capacity middle-mile network is critical, as it carries traffic from multiple last-mile networks and provides service to anchor institutions.

Different broadband technologies can be used to provide this service, including fiber optic, wireless, or satellite backhaul. Fiber optic is usually the technology of choice, as it offers high capacity and reliability in data transmission. The middle mile often operates at a broader scale – such as the county, regional, or state level (Arnold and Sallet, 2020) – but it is not always physically accessible to rural communities. Therefore, rural broadband projects may need to factor-in the costs of building a middle-mile segment.

In addition, many rural communities rely on a single conduit for middle-mile access. They are at risk of losing connectivity if the conduit happens to be damaged due to public works (Ban, 2023). To decrease the risk, states can provide funding for the installation of alternative conduits – thereby making the middle-mile network "redundant." If one conduit is damaged, the community can remain connected through the alternative routes. Network redundancy is a concern of Colorado (Abarinova, 2023), and its Department of Local Affairs awards grants exclusively for middle-mile infrastructure and for planning. We describe how Colorado is supporting middle-mile deployment in Chapter 6 (also see Bravo and Warner, 2024; Bravo, 2025).

Overall, states recognize the criticality of the middle mile for rural connectivity. For ten states, the middle mile is an eligible expense if it serves the last-mile project. The data show that 12% of state grants included middle-mile costs. These combined projects also received a larger grant amount per premise passed ($7,601) than projects that were last-mile-only ($2,226).

Provider Eligibility

States Largely Support for-Profit Providers – But Keep It Local

As with technology, states have the choice of following the traditional path or innovating. States might prefer to fund established, for-profit providers (telephone and cable companies), which tend to have the experience and resources to make expansion quicker and less expensive. They are likelier to own cable or DSL infrastructure near the project area. In these cases, state funds essentially assist with the "extension" of existing infrastructure, rather than the construction of a new fiber optic network, which is a slower and costlier process. This is the case in Alabama, which prioritizes "extension" projects that leverage existing infrastructure. Unsurprisingly, roughly half of Alabama's broadband grants were awarded to cable providers that likely already had the necessary footprint.

Criticism has been directed at federal programs over the large amounts of federal subsidies received by large telecommunications companies (Ali, 2020; Neenan, 2023). In 2014, ten of the largest carriers in the US emerged as the main beneficiaries of the FCC's Connect America Fund Phase II (CAF-II). They were collectively awarded $9 billion to provide service of at least 10/1 Mbps. A year after the awards were announced, the FCC raised the speed benchmark to 25/3 Mbps – but did not change the minimum speed required for CAF-II projects. The outcome is that, until 2021, CAF-II winners have continued to receive FCC funds to deliver 10/1 Mbps. Not long after, the FCC raised the speed benchmark once again – to 100/20 Mbps, BEAD's benchmark (Neenan, 2023). As a result, despite billions in subsidies, the gap inevitably remains. These communities do not qualify as served by the current standards. However, they may not qualify for new funding because federal and state programs are reluctant to subsidize areas that have already been awarded grants.

In contrast, failure to meet coverage and speed expectations has made some states cautious about relying on large telecoms to close the digital divide (Buckley, 2016). States can take a complementary approach and prioritize applications from providers that operate locally or regionally, are headquartered within the state, and primarily serve its population. Providers also can be required to demonstrate their experience. For example, in Alabama and Indiana, Internet Service Providers (ISPs) must operate for a number of years to be eligible. Finally, providers are required to demonstrate their financial capability by covering a significant percentage of the project's costs. Table 3.1 shows that this percentage varies across programs. The higher the match required, the higher the barrier for small providers and communities with limited resources.

The preference for established providers with resources inevitably tilts the balance in favor of "traditional" providers, such as for-profit telephone companies, telephone cooperatives, and cable companies. "Non-traditional" providers, like electric cooperatives and municipal/regional utilities, might only be able to enter the broadband market with the assistance of state funds. There is also the issue of state-level restrictions toward non-traditional providers. As Table 3.1 shows, ten states in our study restricted the ability of local governments to own and/or operate their own broadband networks, thus municipal network providers

TABLE 3.3 State Broadband Grants by Provider Type and Provider Size

Provider Awarded	Provider Size				
	Active in One Region	Multiple Regions or Nationwide	Unknown	Total	
	N	N	N	N	(%)
For-profit telephone company	188	59	0	247	(34)
Telephone cooperative, or a subsidiary	102	0	0	102	(14)
Wireless ISP	80	0	0	80	(11)
Another category	73	6	0	79	(11)
For-profit cable company	12	64	0	76	(10)
Unknown	0	0	60	60	(8)
Electric cooperative, or a subsidiary	57	0	0	57	(8)
Municipality or a municipal utility	23	0	0	23	(3)
Total	**535 (74%)**	**129 (18%)**	**60 (8%)**	**724**	(100)

Source: Pew Charitable Trusts State Broadband Grants 2014–2020, N = 724 funded projects in 17 states.

are not eligible for state funding. We will describe how communities navigate state municipal broadband restrictions in Chapter 7.

All of these considerations impact how state broadband grants are awarded. As Table 3.3 shows, between 2014 and 2020, nationwide telephone and cable companies received only 18% of state broadband grants. States are strengthening the local broadband service market – but most recipients were traditional providers.

This outcome is not exclusively a product of state policy, but also of provider availability. For example, California is mainly served by large regional or nationwide telecoms, rather than small providers (The Pew Charitable Trusts, 2020b). Thus, California primarily funded for-profit telephone companies. In contrast, rural areas in Minnesota and Wisconsin are primarily served by local/regional telephone companies and telephone cooperatives (The Pew Charitable Trusts, 2020b), and they received the majority of grants in both states.

Non-traditional Providers Are Rarely Supported

Only a few awarded projects involved "non-traditional" providers. We describe two types: electric cooperatives and municipal broadband operators.

Electric cooperatives face fewer restrictions than municipal broadband operators, and Table 3.3 shows they received 8% of state grants. Electric cooperatives already deploy fiber optic conduit to upgrade their electric power grids to smart grids. Excess fiber can be leveraged as middle-mile infrastructure to serve rural, last-mile ISPs. Electric cooperatives also have direct access to utility poles and easements, which facilitates installation. They are primarily concentrated in the Midwest and the Southeast (Gilcrease et al., 2022). Several states in our study funded electric cooperatives, including those with the largest number of registered electric cooperatives – Indiana, Iowa, Minnesota, and Missouri. Electric cooperatives received significantly larger grant amounts per premise passed ($5,230) when compared to other provider types ($2,593). This may be due to the fact that most of the grants awarded to electric cooperatives were used to build fiber.

The limited role of municipal broadband operators can be attributed to many factors, including backlash from large telecommunications companies that seek to prevent competitors from entering underserved areas (Muller, 2022; Block, 2022), and state regulation that imposes roadblocks or prohibits municipal entities from owning, financing, and delivering broadband service (Cooper, 2024). Ten out of 17 states in our study restricted municipal broadband in some way. These barriers do not necessarily block municipal broadband operators from state funding, but they could discourage them from applying. For example, Wisconsin restricts municipal broadband, and out of 210 grants only eight were allocated to municipal broadband projects. Private providers argue that entry by municipal providers to underserved markets can oversaturate the market. Still, without competition, there is little incentive for private providers to upgrade or make services more affordable.

In Chapter 7, we show how PPPs are an alternative to get around municipal broadband restrictions. Local governments received more grants when they were part of a public-private partnership (18% of 724 projects) than when applying alone (4%). PPPs were more commonly funded in states that gave weight to PPPs or required grantees to be PPPs, such as Maine and Virginia. While state programs requiring PPPs ensure that local governments will be part of the application process, this does not guarantee public ownership of the network or that it will promote competition. In Virginia, where grantees are required to be PPPs, local governments apply for grants, and their private sector partner owns and operates the network.

Can States Address Digital Equity?

Yes, they can. Before the pandemic, some state broadband programs had already incorporated digital equity concerns into their funding criteria. However, states must take additional steps to reach those communities that are least likely to be served.

State broadband grants often match only a percentage of the project's construction costs, and the rest is to be covered by the private sector partner. While the community might gain additional support from a federal/local grant or loan, the private sector still bears part of the risk. When profitability is factored-in, there is little incentive to build where the return on investment will be low. In rural and low-density areas, broadband networks must expand over long distances to capture as many customers as possible. To recoup costs, monthly charges for new and faster Internet service must increase as well – which might not be profitable in high-poverty areas. In addition, low-income and minority areas are also disproportionately impacted by historic disinvestment in broadband infrastructure – better known as "digital redlining" (Callahan, 2017; Leventoff, 2022).

Overall, there is little incentive to serve rural, high-poverty, and minority areas – with or without grant dollars. Without state grant criteria that explicitly prioritize these communities, providers can use state funds to build in areas that are more profitable. For example, suburban areas are more likely to see private investment than rural areas due to their proximity to urban areas. Indeed, Table 3.2 shows that 37% of awarded projects are located in suburban areas, and 37% in urban areas. While these areas may have a higher proportion of low-income and minority population, their lack of adequate service is likelier due to "digital redlining," than population density.

Table 3.1 shows that some state broadband programs recognized these concerns. We assessed three components of digital equity in state grant criteria: rurality, population density, and poverty. Colorado and Iowa prioritize projects in low-density areas. Alabama requires a percentage of funds be used in unincorporated areas. Colorado's DOLA Broadband Program targets communities impacted by the mineral and mineral fuel industry. Virginia's TRRC Broadband Program funds projects that serve any of their 40 tobacco-dependent counties. California matches an additional percentage of a project's costs in low-income areas. Missouri prioritizes projects in areas with high poverty or unemployment rates. North Carolina and Tennessee prioritize projects in economically distressed counties.

On the other hand, there are grant requirements and state regulations that could become barriers to under-resourced areas and non-traditional providers. The first is a state program's preference to fund "traditional" for-profit providers, which might be uninterested in serving remote and high-poverty areas. Several states restrict electric cooperatives or municipal broadband (Table 3.1), which are often the only options for communities that lack other options.

The second barrier is a program's required match percentage. Many states will require applicants to match a percentage of the project's costs. Table 3.1 shows how broadly the percentage varies across states – from zero in Wisconsin to 85 in Iowa. A higher match can narrow the pool of applicants to providers with resources. Communities may also contribute a local match to increase the application's chances of success. However, a higher match requirement can prevent small and rural providers from participating. Thus, rural and high-poverty communities are left with fewer options.

Access to state broadband grants can be impacted by barriers related to digital equity, local capacity, and state policy. We performed a county-level analysis to explore their impact.[2] We matched state broadband grant data with county-level data on socioeconomic characteristics and provider availability, and with state-level policies (required match percentage and municipal broadband restrictions).[3] These grants were awarded before the pandemic, so our analysis explores the impact of local characteristics and state-level policies before digital equity became a federal policy goal.

Table 3.4 shows a summary of model results. Of the 1,167 counties in the 17 states, 405 counties received at least one grant. We divided the 405 counties by metro status,[4] and found that rural counties received more grants. Among urban and rural counties, places with less population density received more grants as well. This shows that states targeted low-density areas – a key barrier to digital equity. However, other barriers like poverty and race were not targeted. In fact, rural counties with lower poverty received more, not fewer, grants. We suspect that local capacity can prevent communities with fewer resources from applying for state funds. With regard to adoption, rural counties with lower adoption received more grants. Both urban and rural counties with a larger population received more grants, as did rural counties with more educational attainment.

TABLE 3.4 County-Level Differences Related to Digital Equity, Local Capacity, and Policy Barriers

	Which Counties Received more Grants? N = 405 Funded Counties, 17 States	
	Urban (1–9 Grants) N = 158 Counties	Rural (1–13 Grants) N = 247 Counties
Digital equity barriers		
Population density	–	–
Poverty	Not significant	–
Percentage White population	+	+
Adoption rate	Not significant	–
Local capacity barriers		
Total population	+	+
Pop. with a college degree or higher	Not significant	+
Broadband provider availability	Not significant	Not significant
State policy barriers		
Required match %	–	–
Municipal broadband restrictions	Not significant	+

Data source: Author analysis of Pew Charitable Trusts State Broadband Grants 2014–2020.

To our surprise, provider availability did not appear to impact grant alloca-tion. On the other hand, both urban and rural counties in states with lower match requirements received more grants. Finally, broadband restrictions did not have an effect on grant receipt in urban counties, and ironically, had a positive effect in rural counties – possibly because the lack of competition from municipal pro-viders may have encouraged private providers to compete for funds.

While four state programs (Alabama, Colorado, Iowa, and Virginia) specifi-cally prioritized rurality and low population density, we find that population den-sity was significant across all grants. States are easing the costs of building in rural areas, where the rate of return will be lower and service charges higher to recoup the initial investment.

There seems to be no relationship between state grant distribution and other barriers to digital equity. Although Missouri, North Carolina, and Tennessee pri-oritize projects in areas with high poverty and unemployment rates, or that are considered "economically distressed," the overall model shows poverty had no effect in urban areas and a negative effect on grant recipients in rural areas. Model results also show that counties with more grants had a higher percentage of White population. This could be attributed to our sample where half of the grants in our dataset were from Minnesota and Wisconsin – states with a major-ity White population (83% (MN) and 84% (WI), respectively). States will need to pay closer attention to ensure that high-poverty and minority areas receive state broadband funding.

Our results show clearly that gaining access to state broadband funds is also a matter of local capacity. Counties with a larger and more educated popula-tion will have a larger and wealthier tax base. Municipalities in these counties will have more fiscal resources at their disposal to offer a local match, and this increases their chances of receiving a state grant. Providers also will be able to pick areas that are both profitable but still qualify for funding. As Table 3.4 shows, even though states managed to target low-density areas, those with a larger and more educated population received more grants.

Our models also highlight the importance of policy barriers, such as match re-quirements, which narrow the applicant pool to providers with resources, and to communities that are able to contribute with a local match. Other research finds that match requirements are a barrier to rural and high-poverty areas (GAO, 2022; Smith, 2023). Capacity is a constraint, and states will need a more flex-ible approach to bolster participation by small and rural providers. For example, California has a sliding scale to encourage deployment in unserved areas: it does not require a match for projects in unserved areas, but does require a 40% match for projects in underserved areas.

Our models found that in states with municipal broadband restrictions, rural counties received more state grants. This was a surprise. If could be that munici-palities in these counties may lack the capacity to finance their own networks in the first place, and so state-level restrictions may have little impact. It is also possible that the guarantee of less competition (from municipal providers) may facilitate private provider interest in some rural counties. In Chapters 4 through 7 we discuss how communities explore alternative PPPs, especially in states with municipal broadband restrictions (also see Bravo and Warner, 2024).

The Path Forward for State Broadband Policy

In some ways, state broadband programs have laid the groundwork for broader notions of access and equity. Despite being technology neutral, they were primarily funding fiber optic infrastructure before it became a federal priority. They supported local providers, rather than nationwide telecommunications and cable companies. Some states already had some digital equity concerns embedded in their funding criteria, including rurality and low population density. Then federal policy followed, as the pandemic highlighted ongoing socioeconomic disparities in broadband access and adoption.

The path forward requires states to reassess their role in subsidizing the market and supporting non-traditional providers. The implicit preference for private, for-profit providers could limit options in rural and high-poverty areas. Where state funding may fail to attract private investment, communities may want to explore partnerships with non-traditional providers. However, electric cooperatives and municipal broadband operators face regulatory restrictions and receive only a small percentage of state grants. State-level policies which limit the range of players are restricting competition in the most underserved markets.

State programs are poised to receive billions from the BEAD Program. The broader definitions of digital equity, and encouragement of alternative municipal and non-profit providers will challenge states to broaden their program rules. We have shown the extent to which prior state broadband policy, considered digital equity barriers, and local capacity constraints in their funding allocations. This is a lesson taught by states themselves, some of which were already prioritizing projects in rural and economically distressed areas.

Future state program design must address broader digital equity barriers and capacity constraints. A narrow focus on unserved and underserved locations, without weighing in the factors that contribute to poor connectivity in these locations, may contribute to disparities in the distribution of state broadband grants. Rurality and low population density are only two barriers to digital equity. Disparities in access and adoption are linked to other demographic and socioeconomic factors as well – including poverty, race, and education. States will need to carefully consider how specific eligibility criteria, such as prohibitive match requirements, can become barriers to participation by communities with the greatest need.

Notes

1 BEAD was restructured by the Trump Administration in June 2025. This chapter references elements of BEAD in place during the Biden Administration.
2 Our primary challenge was the limited information on location. Our analysis is at the county level, but applicants could be counties, municipalities or private providers. County capacity could have an impact on how state broadband grants are distributed. Later chapters in this book will show how providers often look to counties for a match contribution. For further detail on the methodology, see Bravo and Warner (2023) and Bravo (2025).
3 We matched the grant data with metrics on metro status, socioeconomic characteristics from the American Community Survey (2016–2020); county-level data on retail provider availability from the Federal Communications Commission; and state policies regarding required match % and municipal broadband restrictions. For further detail on the methodology, see Bravo and Warner (2023) and Bravo (2025).
4 Using the number of grants by county as a dependent variable, we ran a Poisson regression with standard errors clustered at the State-level. For further detail on the methodology, see Bravo and Warner (2023) and Bravo (2025).

Bibliography

Alabama Broadband Accessibility Act, Al. Code § 41-23-210 (2023). https://law.justia. com/codes/alabama/title-41/chapter-23/article-13/section-41-23-210/

Ali, C. (2020). The Politics of Good Enough: Rural Broadband and Policy Failure in the United States. *International Journal of Communication* 14, 5982–6004. https://ijoc. org/index.php/ijoc/article/view/15203

Ali, C. (2021). *Farm Fresh Broadband: The Politics of Rural Connectivity*. Information Policy Series. Cambridge, MA: The MIT Press. 306 pp. https://doi.org/10.7551/ mitpress/12822.001.0001

Ali, C., Berman, D. E., Forde, S. L., Meinrath, S. and Pickard, V. (2024, May 16). The Bad Business of BEAD. Benton Institute for Broadband & Society. https://www. benton.org/blog/bad-business-bead

Abarinova, M. (2023, January 16). Colorado Broadband Chief Talks Local Deployment Challenges, Funding Ops. Fierce Network. https://www.fierce-network.com/broadband/ colorado-state-broadband-official-talks-local-deployment-challenges-funding-ops

Abarinova, M. (2024, August 7). Satellite Broadband Joins the Party for BEAD: What You Need to Know. Fierce Network. https://www.fierce-network.com/broadband/ satellite-broadband-joins-party-bead-what-you-need-know

Afflerbach, A. (2022). Fixed Wireless Technologies and Their Suitability for Broadband Delivery. Benton Institute for Broadband & Society. Retrieved from: https://www. benton.org/sites/default/files/FixedWireless.pdf

Arnold, J. and Sallet, J. (2020) If We Build Them, Will They Come? Lessons from Open-Access, Middle-Mile Networks. Benton Institute for Broadband & Society. https://www.benton.org/publications/middle-mile

Ban, C. (2023, November 6). Middle Mile Can Be a Matter of Life and Depth. National Association of Counties. https://www.naco.org/news/middle-mile-can-be-matter-life- and-depth

Beard, T. R., Ford, G. S., Spiwak, L. J. and Stern, M. (2020). The Law and Economics of Municipal Broadband. *Federal Communications Law Journal* 73, 1–98. https://papers.ssrn.com/sol3/papers.cfm?abstract_id=3819753

Block, B. (2022). How Comcast and Other Telecoms Scuttle Rural WA Broadband Efforts. Crosscut. Retrieved from: https://crosscut.com/news/2022/03/how-comcast-and-other-telecoms-scuttle-rural-wa-broadband-efforts

Bravo, N. (2025). Multi-level Governance for Broadband Planning: Implications for State Policy and Interlocal Cooperation. Unpublished PhD dissertation, Cornell University, Ithaca, NY.

Bravo, N. and Warner, M. E. (2023). Closing the Broadband Infrastructure Gap: State Grant Funds and the Digital Divide. Department of City and Regional Planning, Cornell University, Ithaca, NY. https://labs.aap.cornell.edu/node/880

Bravo, N. and Warner, M. E. (2024). Innovative State Strategies for Rural Broadband: Case studies from Colorado, Minnesota and Maine. Department of City and Regional Planning, Cornell University, Ithaca, NY. https://labs.aap.cornell.edu/node/882

Buckley, S. (2016). Frontier Sued for Alleged Misuse of $40.5M in Federal Broadband Stimulus Funds. Fierce Network. Retrieved from: https://www.fierce-network.com/telecom/frontier-sued-for-alleged-misuse-40-5m-federal-broadband-stimulus-funds

Business Oregon (n.d.) Rural Broadband Capacity Program. https://www.oregon.gov/biz/aboutus/boards/bac/pages/rural_broadband_capacity_pilot_program.aspx

Callahan (2017, March 10) AT&T's Digital Redlining of Cleveland. NDIA. https://www.digitalinclusion.org/blog/2017/03/10/atts-digital-redlining-of-cleveland/

California Public Utilities Act, Ca. Pub Util Code § 281 (2023). https://law.justia.com/codes/california/code-puc/division-1/part-1/chapter-1-5/section-281/

ConnectMaine. (2020). State of Maine Broadband Action Plan. https://www.maine.gov/connectme/sites/maine.gov.connectme/files/inline-files/Plan_Action_2020.pdf

Connected Nation. (2023). Bead and the Choice of Technology. https://connectednation.org/press-releases/bead-and-the-choice-of-technology

Congressional Research Service. (2011). Broadband Infrastructure Programs in the American Recovery and Reinvestment Act [Report]. https://crsreports.congress.gov/product/pdf/R/R40436/26

Colorado Broadband Office (n.d.). Broadband Deployment Board and Fund. https://broadband.colorado.gov/broadband-deployment-board-fund

Colorado Broadband Office (n.d.). DOLA Broadband Program. https://broadband.colorado.gov/dola-broadband-program

Cooper, T. (2024, September 17). Municipal Broadband Remains Roadblocked In 16 States. BroadbandNow. https://broadbandnow.com/report/municipal-broadband-roadblocks

Dano, M. (2022). WISPs Freak Over Guidelines that may Exclude FWA in Unlicensed Spectrum. Light Reading. https://www.lightreading.com/digital-divide/wisps-freak-over-guidelines-that-may-exclude-fwa-in-unlicensed-spectrum

Dawson, D. (2023, February 15). Let's Stop Talking About Technology Neutrality. POTs and PANs. https://potsandpansbyccg.com/2023/02/15/lets-stop-talking-about-technology-neutral/

GAO. (2022, August 11). Breaking Down Barriers to Broadband Access. https://www.gao.gov/blog/breaking-down-barriers-broadband-access

Gilcrease, W., DiCosmo, V. and Padovan, D. (2022). Trends of Rural Electric Cooperatives in the United States from 1990 to 2019: An Empirical Analysis. *Renewable and Sustainable Energy Reviews* 166, 112641. https://doi.org/10.1016/j.rser.2022.112641.

Indiana Office of Community & Rural Affairs. (n.d.). Next Level Connections Broadband Grant Program. https://www.in.gov/ocra/nlc/

Internet for All. (n.d.). BEAD Allocation Methodology. Retrieved on June 3, 2024 from: https://www.internetforall.gov/program/broadband-equity-access-and-deployment-bead-program/bead-allocation-methodology

Leventoff, J. (2022, June 10). Digital Redlining: Why Some Older Adults Overpay for Bad Internet. National Council on Aging. https://www.ncoa.org/article/digital-redlining-why-some-older-adults-overpay-for-bad-internet

Lichtenberg, S. (2017). Broadband Availability and Adoption: A State Perspective. National Regulatory Research Institute. https://pubs.naruc.org/pub/FA869FEA-FDDE-7070-3340-DE38CB16DFC2

Massachusetts Broadband Institute. (n.d.). Broadband Extension Program. https://broadband.masstech.org/about-mbi/past-programs-legislation/broadband-extension-program

Massachusetts Broadband Institute. (n.d.). Last Mile Program. https://broadband.masstech.org/last-mile-programs/last-mile-program-policy

Minnesota Employment and Economic Development Office. (n.d.). Broadband Grant Program. https://mn.gov/deed/programs-services/broadband/grant-program/

Missouri Department of Economic Development. (n.d.). Broadband Grant Program. https://ded.mo.gov/programs/business-community/missouri-broadband-grant-program

Muller, W. (2022). Telecom Giant Moves to Stop Broadband Grant for Northeast Louisiana. *Louisiana Illuminator*. https://lailluminator.com/2022/09/01/telecom-giant-moves-to-stop-federal-broadband-grant-for-northeast-louisiana/

Neenan, J. (2023, November 9). 'It Was Graft': How the FCC's CAF II Program Became a Money Sink. Broadband Breakfast. https://broadbandbreakfast.com/it-was-graft-how-the-fccs-caf-ii-program-became-a-money-sink/

NTIA. (2022). Notice of Funding Opportunity. Broadband Equity, Access and Deployment Program. https://broadbandusa.ntia.doc.gov/sites/default/files/2022-05/BEAD%20NOFO.pdf

North Carolina Department of Information Technology. (n.d.). GREAT Program. https://www.ncbroadband.gov/documents/great-grant/management/great-state-grant-guidance-doc-round-1/

Office of the Chief Information Officer of the State of Iowa. (n.d.). Notice of Funding Availability. https://dom.iowa.gov/media/342/download?inline

S.1167-116th Congress (2019–2020): Digital Equity Act of 2019. (2019, April 11). https://www.congress.gov/bill/116th-congress/senate-bill/1167/text

Smith, N. (2023). High-Cost Areas, Match Waivers, and the Problem of Commercial Sustainability. *Connected Nation*. https://connectednation.org/blog/2023/03/14/high-cost-areas-match-waivers-and-the-problem-of-commercial-sustainability/

Tennessee Department of Economic & Community Development. (n.d.). Tennessee Broadband Accessibility Grant. https://www.tn.gov/ecd/rural-development/tennessee-broadband-grant-initiative/tnecd-broadband-accessibility-grant.html

Temple University Center for Public Health Law Research. (2019). State Preemption Laws. https://lawatlas.org/datasets/preemption-project

The Pew Charitable Trusts. (2020a, November 16). States Tap Federal CARES Act to Expand Broadband. https://www.pewtrusts.org/en/research-and-analysis/issue-briefs/2020/11/states-tap-federal-cares-act-to-expand-broadband

The Pew Charitable Trusts. (2020b, February 27). How States Are Expanding Broadband Access. https://www.pewtrusts.org/en/research-and-analysis/reports/2020/02/how-states-are-expanding-broadband-access

The Pew Charitable Trusts. (2021). How State Grants Support Broadband Deployment. https://www.pewtrusts.org/en/research-and-analysis/issue-briefs/2021/12/how-state-grants-support-broadband-deployment

Vermont Department of Public Service. (2016). Request for Proposals: Extension of Broadband Service to the Underserved Locations throughout Vermont. https://publicservice.vermont.gov/sites/dps/files/documents/Connectivity/RFP_2016_R2/ConnectivityInitiative_2016_R2_RFP.pdf

Virginia Telecommunications Initiative. (2023). Program Guidelines and Criteria. https://www.dhcd.virginia.gov/sites/default/files/Docx/vati/public-input-draft-2023-vati-guidelines-and-criteria.pdf

Virginia Tobacco Region Revitalization Commission. (2017). Last Mile Broadband Program: Program Guidelines. https://revitalizeva.org/wp-content/uploads/2017/08/Last-Mile-Program-Guidelines-vs.4-final.pdf

Washington State Department of Commerce. (n.d.). CERB Rural Broadband. https://www.commerce.wa.gov/building-infrastructure/community-economic-revitalization-board/rural-broadband/

Wisconsin Broadband expansion grant program; Broadband Forward! community certification, Wi. Stat § 196.504 (2020). https://law.justia.com/codes/wisconsin/2020/chapter-196/section-196-504/

PART 2
Local Initiatives Lead the Way

This section takes a closer look at local case studies in several states, and focuses on local leadership in broadband expansion. We lift up the voices of active agents at the local level. These are the people with the vision, persistence, and collaboration skills to build partnerships and address the digital divide. We celebrate their success and identify challenges they face in negotiating market and policy challenges. We present a framework for local action and provide case studies from communities across the US. These include efforts in California, Colorado, Florida, Minnesota, Maine, Pennsylvania, Texas, and Virginia. One challenge of a multi-level governance framework is that states can limit local leadership. Chapter 7 specifically addresses how localities form public-private partnerships to get around state preemption of municipal broadband.

Chapter 4 presents a framework for local action, illustrated with case studies in California and Texas. We then look at three states, and explore the challenges of delivering high-capacity broadband infrastructure in rural communities. Chapter 5 focuses on Minnesota, which operates one of the longest-running broadband programs in the country, and primarily funds local/regional telephone companies and telephone cooperatives. Chapter 6 focuses on Colorado and Maine. Colorado is making last-mile rural connectivity more reliable by providing a strong foundation of high-capacity, middle-mile fiber optic networks. Electric cooperatives are emerging as key players in this field, as they can leverage their own fiber and access to utility poles and electric easements. Maine

DOI: 10.4324/9781003619208-5

allows municipalities to enter interlocal agreements, "Broadband Utility Districts," with their neighbors, and share the costs of deployment and operation. Finally, Chapter 7 explores how communities operate in states with more restrictive broadband policies by partnering with non-profit Internet Service Providers. Together, the chapters in this section showcase our multi-level governance and community resilience frameworks in action.

4

FRAMEWORK FOR LOCAL ACTION

Y. Edward Guo and Mildred E. Warner

Local actors at the municipal and county level play a significant role in advancing broadband deployment. Despite the Internet's ability to connect information from all around the world, broadband is local at the fundamental level: users and providers are local, and the underlying broadband infrastructure depends on technologies that are location-dependent (Ali 2021). Wi-Fi, cellular, and other forms of wireless broadband technologies still require fiber or cable backhaul, meaning the quality and capacity of local, physical infrastructure remain important in determining users' connectivity quality (Sylvain 2012; Ali 2021; Grubesic and Mack 2015). In America's fractured broadband landscape, physical location has a large impact on access to broadband. Our multi-level governance framework, presented in Chapter 1, emphasizes the critical importance of local leadership. In this chapter we articulate a local framework for action.

Gillett et al. (2004) identify four roles municipalities can play in the broadband development process. Municipal governments can be key broadband users and use their leadership role to "assess, stimulate or aggregate demand." Local governments can measure demand by conducting surveys, stimulate demand by training businesses and citizens, and aggregate demand by using the government itself as an anchor tenant to attract providers. Local government can act as a rule-maker and adopt or reform local ordinances to facilitate broadband development. Municipalities can expand access to local facilities such as utility poles or coordinate better planning. Local governments also can act as a financier, providing economic support to users or providers. Such support can take the form of grants, loans, or tax incentives for providers, or of devices and services for customers. Additionally, the local government itself can act as an infrastructure developer and construct and operate municipal networks.

DOI: 10.4324/9781003619208-6

Local governments and local institutions have played all the above roles in communities across the US. Local leaders understand the implementation challenges to broadband delivery, and have implemented numerous innovative solutions. This chapter examines the broadband initiatives of two communities, the City of Brownsville, TX and Santa Cruz County, CA. Using a community resilience framework, we find four elements matter: the development and engagement of community resources, the presence of active agents, a project's positive impact, and the opportunities created by the pandemic (Magis, 2010; Ashmore et al., 2017; Roberts et al., 2017). The case studies show the potential of public-private partnerships (PPPs) when there is a shared mission to address equity.

Community Resilience and Broadband Delivery

Resilience is a conceptual framework to examine how social systems adapt to change (Roberts et al., 2017). While many definitions of community resilience exist, the most complete definition comes from Magis (2010) and its extension by others (Ashmore et al., 2017; Fawcett et al., 1995; Roberts et al., 2017). According to Magis (2010), community resilience is the existence, development, and engagement of community resources to thrive in an environment characterized by change, uncertainty, and surprise. Resilient communities intentionally develop a personal and collective capacity to respond to and influence change, sustain and renew the community, and develop new trajectories for the future (Magis, 2010).

The resilience framework represents the foundation for building digital access. However, there is more to digital equity than simply building access. That is only the first step. It is also important to address affordability and adoption. These three As – access, affordability, and adoption – are the components of digital equity (Gonsalves, 2021). The digital opportunity compass takes governance and capacity as its foundation and then moves on to imagine how adoption can lead to real changes in health, education, and economic outcomes (Rhinesmith et al., 2023). Communities imagine the future – when digital equity is present – to begin working on how equity will be embedded in broader community education, employment, and health systems. Communities also must consider that, for many residents, cell phones and Internet hotspots will be the main tools to access the Internet. To center equity and adoption, careful attention must be given to the needs, training, usage patterns of the poor, elderly, and other marginalized groups (Turner Lee, 2024).

The community resilience framework helps us explore how communities respond to disruptions and market failure. Resilience is a dynamic process, and in broadband, the situation is constantly changing. The COVID-19 pandemic helped focus attention on broadband. Communities do not control all the factors that affect them, so resilience may depend on factors originating beyond the community itself. New technologies, standards, maps, and government policies can all impact project development. Broadband failure in the US is the result of market and policy failures, as described in Part 1 of this book. These factors are outside the control of individual communities but define the limits of local broadband projects.

To evaluate community resilience, Magis (2010) emphasizes: community resources, active agents, collective and strategic action, equity, and impact. Community resources encompass a wide range of natural, human, social, financial/built, and political capital (Flora et al., 2018). Building up and actively utilizing community resources is mutually reinforcing. Active agents influence decisions and take the lead. Through collective action the community develops partnerships for action. Strategic action encompasses the process of debate, planning, implementation, and learning within the community. Equity addresses the challenges of equal access and distribution of costs and benefits, while impact measures the success of project implementation.

We incorporate additional analysis elements in our resilience framework to provide greater clarity on how active agents and collective action enhance resilience. Doussard and Schrock (2022) argue that urban policy entrepreneurs, as active agents, can effect change through a combination of building power with outside partners, utilizing data and expertise, and working with local government to advocate for successful solutions. The core relationships that affect collective action include reputation, trust, and reciprocity, which combine to increase cooperation (Ostrom, 2010). Collective action is enhanced by sharing a common agenda, common metrics, mutually reinforcing activities, continuous communication, and a backbone support organization (Kania and Kramer, 2011).

Resilience is multi-dimensional (Roberts et al., 2017). Power issues and relationships must be considered to ensure that the resilience of the entire community is truly enhanced (Ashmore et al., 2017; Ashton and Kelly, 2019). The unexpected can also have a major impact on resilience. The COVID-19 pandemic highlighted the need to address broadband inequity, and the resulting American Rescue Plan Act (ARPA) provided additional resources. Both case studies presented in this chapter seized upon the opportunity and moved quickly during the pandemic to launch new broadband equity programs. Addressing the digital divide requires attention to redistributional, procedural, and conceptual equity. Political will and partnership building have been key to communities addressing equity in case studies of community broadband since the COVID-19 pandemic (Diaz Torres and Warner, 2024).

In this chapter, we present two case studies, Brownsville, Texas and Santa Cruz, California. Key stakeholders were interviewed and related news reports and government documents provided broader context. Interviews explored why and how stakeholders chose to engage in broadband activities; how different partnerships were established; the status of the project; how projects are funded, financed, managed, and operated; how challenges encountered were resolved; and how projects are ensuring long-term program sustainability. Additional details can be found in Guo (2023).

Our interviewees represent the key actors in the coalitions that made broadband development possible in both communities. In Brownsville we interviewed Elizabeth Walker, Assistant City Manager and Marina Zolezzi, Chief of Staff for the City; Rene Gonzalez. Chief Policy and Compliance office at Lit Communities (the ISP), and Jordana Barton-Garcia, former community development banker and Senior Advisor for the Federal Reserve Bank of Dallas. In Santa Cruz County, we interviewed Jason Borgen, Chief Technology and Innovations Officer at Santa Cruz County Office of Education (COE); James Hackett, COO at Cruzio Internet (the ISP); and Kevin Heuer, Director of Engagement & Impact at Community Foundation Santa Cruz County.

Brownsville, TX

Brownsville, TX, has long suffered from subpar broadband connections. Over the past decade, it has appeared multiple times on the National Digital Inclusion Alliance's "Worst Connected Cities" list and topped it twice (NDIA, 2019; NDIA, 2018; NDIA, 2017). By 2019, the city was motivated to rid itself of this infamous label, and disruption from COVID-19 intensified Brownville's efforts. Using $19.5 million of its American Rescue Plan Act (ARPA) funding, the city entered into a PPP with Lit Communities, which is committed to providing an additional $70 million, to construct a publicly owned middle-mile network connecting last-mile local networks to high-capacity national and regional backhaul and connect everyone within the city limits. When finished, Brownsville will have 100 miles of public open-access fiber middle-mile backbone and 550 miles of private last-mile fiber connections to homes and businesses (Treacy, 2022).

The Problem: Histories of Poverty and Private Market Inefficiencies

Brownsville's lack of broadband stems from the city's legacies of poverty and market failure in broadband. Founded on cycles of violence against Mexican and native residents in the 19th Century, Brownsville was the result of "repeated waves of Spanish, Mexican, Texan, and American colonization for economic gains" (MacWillie et al., 2021). The local economy, based on trade and agribusiness, has failed to flourish, resulting in the city's frequent appearances in "America's poorest cities" rankings (MacWillie et al., 2021; Hlavaty, 2013; De-Pietro, 2021). Residents often face insufficient and volatile income streams, lack access to credit and financial institutions, and lack health and retirement benefits (Diaz-Pineda, 2021). The Colonia communities surrounding Brownsville, consist of mostly substandard housing, without basic infrastructure (Barton et al.,

2015). Legacies of poverty meant that there "simply wasn't [enough profit motive] for [private ISPs] in the Brownsville marketplace," says Walker, causing a market failure that has left Brownsville's Internet needs unmet.

Internet connection in Brownsville is deeply inadequate. A city-sponsored survey in 2022 showed that 66% of the city's residents lacked DSL, cable, or fiber access, and 23% had no broadband access of any kind, including cellular data plans; additionally, 32% of residents had connection speeds below the 25/3 FCC standards, and 65% had download speeds below 100 Mbps (Lit Communities, 2022). The American Community Survey similarly shows that only one census tract around Brownsville has fiber, cable, or DSL subscription rates higher than the Texas state rate of 68.77%, with the lowest census tract at only 9.57% (U.S. Census Bureau, 2021). Interviews with city officials show that local emergency departments sometimes report difficulties communicating with each other due to poor network connection, and a Brooking's report found that companies have left the Brownsville region due to inadequate levels of broadband connectivity that constrained business growth (Tomer et al., 2020).

Getting Started: We Just Decided to Move

Although Brownsville's broadband issues had been known for a decade, it took real determination from community digital champions to get improvements underway in 2019. The first champion was the newly elected mayor. After seeing the city named one of the nation's "Worst Connected Cities," he was determined to act, as he knew that for Brownsville "to be a thriving community, the issue of broadband must be one of [his] top priorities" as he wrote in an opinion piece (Mendez 2022). In his 2019 State of the City address, he stated that

> access to Broadband will determine [the city's] future … [and he had] heard many stories that [students], the future leaders of [the] community, struggle with not having access to the Internet, whether it's because of affordability or just lack of infrastructure to their homes. (Barton 2021)

He called the situation "unacceptable," and quickly got to work to rectify it.

Another important community digital champion was a community development banker and senior advisor for the Federal Reserve Bank of Dallas, Jordana Barton-Garcia, who city officials referred to as the "original instigator" of Brownsville's broadband project (Cameron, 2022). During her research into infrastructure in the border Colonias in the early 2010s, Barton-Garcia made a surprising discovery about the importance of broadband. Even though broadband was not included in any questions asked during the research, residents in the Colonias shared countless stories of how limited Internet access was constraining children's schoolwork and residents' ability to participate in the job market (Barton, 2021; Barton et al., 2015). Realizing broadband's critical role in economic development, she made it her focus to learn about the root causes and practical solutions to the digital divide. In the process she managed to build relationships in the field with communities, policymakers, and providers.

The two community champions combined forces after the mayor's inauguration. The mayor began organizing key community stakeholders in Brownsville, including economic development agencies, the local community college, the school district, and local utilities, to meet and discuss solutions, and the community development banker was invited to join (Treacy, 2022). Barton-Garcia gave presentations on best practices and options based on her prior work with the City of Pharr, TX, and other communities in their process to improve broadband. Further, she introduced the city to organizations that could offer help, including Brownsville's eventual PPP partner, Lit Communities.

The COVID-19 pandemic deepened the political realization of broadband's importance and hastened Brownsville's action. During the worst pandemic waves, the lack of reliable Internet connectivity significantly hampered the Brownsville economy and the daily lives of its residents. The city found that close to 70% of its labor force could not work because connections were not fast enough to sustain remote work. Children could not access school materials. People struggled to access telehealth. COVID-19, according to Walker, "really laid very plain, raw and bare exactly what are the consequences born by a community that is digitally disconnected."

In summer 2020, the city started planning for broadband improvement first by conducting a survey to understand the existing broadband services. The survey showed that many residents were paying high prices for substandard services, and that many fundamentally lacked access to affordable, high-speed broadband (Lit Communities, 2022).

Project Development: Partnerships and ARPA

Information provided by the survey led the city to conclude that to truly address its broadband deficiencies, it needed to connect every home and business with fiber. However, when the city costed out the entire plan, it realized it lacked funds to complete the project. According to city officials, the city had initially considered debt financing or rate recapture via the municipal electric utility to pay for broadband construction, but this would have delayed progress on the project. When the city was allocated over $60 million in ARPA funding in 2021, the city suddenly found itself with the ability and resources to "make [its] own investment, … have an ownership stake, … and control [its] own destiny," says Walker ("Cultural and Tourism Grant Fund" 2022). During this process, the city administration had multiple rounds of discussion with the city commissioners regarding the best approach for the project. In the end, after weighing all costs and benefits, the city commissioners decided that a PPP model would be the best approach for Brownsville. The city would use $19.5 million of its ARPA funding, made available in early 2021, to construct an open-access middle-mile network, and would enlist the help of private providers to connect to homes and businesses. Reflecting on the decision, Walker said,

> [By utilizing the PPP model], we had some skin in the game. And I think that was important to our Commission too. That way, we would be able to assert ourselves and assert the role of the city and its responsibility to the community. Our commission felt very strongly that there had been a market failure by the capitalist approach towards private investment. So they want to be able to have some skin in the game, and they thought by owning the middle mile, that would be a way to hold accountable whoever would be the purveyor of the last mile. Now we are going forward, and it also meant that we could then create a competitive marketplace, because then we could open [the middle-mile network] up to other ISPs for their participation.

With the PPP approach decided, the city sent out Requests for Proposals in search of a private partner. The city collected 20 responses, with no bid from the traditional ISPs. The city administration engaged in conversation with eight providers and shortlisted four. After a series of negotiations, the city eventually settled on a partnership with Lit Communities (hereafter Lit). Brownsville's $19.5 million contribution was able to leverage a $70 million commitment from Lit (Treacy, 2022). Under the agreement, the city's money will pay for a publicly owned open-access middle-mile fiber network that can generate revenue for Brownsville via leases to potential ISPs, and Lit's portion will help establish private fiber-to-the-home connections to all residents and businesses. In addition to constructing both the middle-mile and last-mile networks, Lit will be responsible for operating and maintaining the consumer-facing network.

Reflecting on the entire project development process, city officials believe that ARPA money really helped propel the project forward, calling the federal funding "a once-in-a-lifetime opportunity to make a singular investment that could have transformational change" for Brownsville. Without the ARPA funding as an initial investment, the city would not have been able to leverage as much private funding and start construction this quickly. The project would have taken many years to realize.

Benefits of PPP for Everyone

The partnership between Lit and Brownsville has many benefits to both parties and the city's residents. First, the PPP model allowed the city to increase its return on investment. With $19 million, Brownsville leveraged an additional

$70 million in private investment, more than tripling the amount of capital the city could provide on its own. Second, the city will be able to earn revenue and promote competition with its open-access middle-mile network. Future providers wishing to use the municipal middle-mile network to expand their customer base will need to pay Brownsville a user fee. The open-access nature of the network should help increase competition and lower broadband prices in Brownsville because such networks lower the infrastructure cost for new market entrants.

When asked why Lit would willingly agree to constructing and maintaining an open-access network, Gonzalez said that Lit's model in Brownsville "only needs roughly around half of the serviceable locations to give the take rate [needed] to give the ROI that is acceptable to not only [Lit] but [its] investors," and its survey show that 93% of residents would register if enhanced broadband services were made available (Lit Communities, 2022). Third, public utility operators are able to leverage the broadband infrastructure to modernize their infrastructure as the City and Lit collaborated on route planning. City Hall, police stations, and other community institutions will be connected, and the local public utility is building new automatic metering technology from the broadband network. Fourth, Lit is committed to training a cohort of residents to work on installing and maintaining the network, creating new job opportunities in the area. "What better people to take care of a network than its own citizens and its own people from its own community?" noted Gonzalez. Fifth, the partnership agreement between Lit and the City guarantees that lower-income individuals will receive 100/100 speeds capped at $30 per month even after the federally funded Affordable Connectivity Program (ACP) was discontinued, as Lit is committed to working with local non-profits to continue the program. The city, tagging onto Lit's effort, also plans to step up its work on enhancing digital equity in general by introducing device provision programs with public and private institutions.

One additional benefit of Brownsville's PPP model is that it ensures the city will not face hurdles with regard to state preemption of municipal broadband (Casper, 2021). The Texas municipal broadband prohibition was less than absolute after the city of Mont Belvieu obtained permission from state courts to build and operate its municipal broadband network (Silverman, 2019). Brownsville had followed the case closely and was aware of the possibilities of direct public provision. In fact, Brownsville even applied parts of the Mont Belvieu ruling to its own advantage when it was planning its broadband project. Initial strategizing with the city commission, as the ruling allowed, happened behind closed doors, allowing the city to keep its cards close before it was ready to make any public announcement. More importantly, Brownsville will remain unaffected by the apparent preemption laws because the partnership with Lit means the city is not directly providing broadband and therefore does not fall under the scope of the state preemption. Additionally, the city simply saw more advantage in involving private entities in the process compared to direct public provision, as the former approach allows the city to reduce maintenance and customer service responsibilities while promoting competition.

Hurdles: Incumbent Opposition

A surprising barrier to Brownsville's ambitious broadband improvement project was opposition from the incumbent, large ISPs. Despite not participating in the PPP bidding process or promising service expansions, existing ISPs made various attempts to stall the project. Leading the opposition were AT&T and Charter Communications. An AT&T press release, first reported by the *Rio Grande Guardian*, argued that "some local officials in the Rio Grande Valley are pushing to spend federal funds to build government-owned networks that would connect only public and government buildings – not households," and that "networks owned by local governments often fail due to lack of expertise and money, leaving taxpayers responsible for millions of dollars of debt" (Taylor, 2022b). AT&T's Vice President for External Affairs, J.D. Salinas, further stated to *Rio Grande Guardian* that Brownsville's approach hampers security,

> What's important to know is patching together a network with multiple providers presents operational risk, cybersecurity risk, and continuity issues. AT&T is committed to building, operating, maintaining and upgrading our networks so that our customers have high quality and secure experiences. We cannot maintain that standard of excellence when utilizing a middle-mile network maintained by another provider. Quality and security may be compromised when a network is pieced together in a middle-mile scenario (Taylor, 2022b).

Charter made similar comments according to reporting by *Rio Grande Guardian*, with the communication company's Vice President for State Government Affairs Todd Baxter arguing that "Brownsville is very well served and ubiquitously served by the private sector," and questioned whether it is "a good use of taxpayer dollars" to overbuild the private sector (Taylor, 2022a).

Beyond criticisms, incumbents paid for advertising campaigns to boast about their services. Incumbent providers went as far as enlisting the local chapter of the Council for Citizens Against Government Waste to file Freedom of Information Act requests demanding the release of Lit's proprietary business model. Such efforts were unable to derail the city's broadband project.

City officials argue the incumbents' claim that Brownsville's model is built only for public purposes is untrue as it has always been the city's goal to connect every premise within city limits. The city also had survey data to show that incumbent providers do not deliver affordable, reliable broadband services to

Brownsville residents. In the end, the mayor delivered the best response to the incumbent opposition via the *Rio Grande Guardian*,

> AT&T had their chance. Spectrum had their chance. All they did was try to prevent us from connecting ourselves as a community. We did not find they were willing to come to the table with any sort of solution. It was more about trying to convince us that everything was okay when obviously it was not. So, we are moving forward (Taylor, 2022b).

Santa Cruz County, CA

Santa Cruz County, CA has long endured substandard Internet connections, especially in its rural areas. When COVID wreaked havoc on the county's education system, community organizations – including local school districts, the local Community Foundation, and a local ISP, Cruzio, built on each other's strength and delivered fast broadband connection to some of the county's poorest and most isolated communities. Hundreds of thousands of dollars' worth of donations, grant funding, and Cruzio's own match have made the project a reality that has continued to expand.

Challenges: Geography, Demographics, and Market Failure

Santa Cruz County is unique in its geography. The southern end of the county sits along the California coastline and is dotted with residential and farming communities. The northern end of the county, however, is occupied by the Santa Cruz mountains and covered by rocks, valleys, and other challenging terrains. This terrain, coupled with lower population densities in northern Santa Cruz County, makes it difficult for Internet providers to build high-speed connections at a profit. According to Hackett, areas in the "Santa Cruz Mountains…were until recently still reliant on an old DSL service, which has recently been pretty much discontinued. And now pockets of [the county's] populations are left with no option other than satellite service or cell service."

Terrain and geography, however, are not the only roadblocks to Internet access in Santa Cruz County. Unlike northern Santa Cruz which has more ties with Silicon Valley, southern portions of the county are dedicated to agriculture, and host migrant farmworkers, often struggling with financial independence and living in remote farming communities, and thus do not present strong incentives for major ISPs to enhance broadband service. Around 60% of students in the county's education system are English language learners, and a similar percentage are socioeconomically disadvantaged. In Pajaro Valley Unified School District, which serves around half of the county's students whose parents are farmworkers, 79.2% of students are socioeconomically disadvantaged, and 40.3% are

English language learners (Monroy, 2021). An example of how the broadband needs of students and their families are not met is Buena Vista Migrant Center, home to 103 farmworkers and families in southern Santa Cruz, which has no Internet access. One resident recalls that AT&T offered DSL connections, but the service was discontinued without clear explanation (Monroy, 2021).

Impetus: The Pandemic Changed Everything

While access and affordability issues long existed in Santa Cruz and some attempts were made to address the issue, including providing devices in local libraries and schools and conducting a broadband survey; it was the outbreak of the COVID-19 pandemic that finally rallied the community to examine the broadband situation in detail and venture for more systemic solutions (Dolgenos et al., 2020; Isenberg, 2018).

The pandemic, and its negative impact on children's education, was the most important driver for action in Santa Cruz County. When the pandemic hit, the county's Office of Education (hereafter COE) realized that "all of a sudden from one week to the next, all the kids, all the classes are going online and all the kids need to have a broadband connection," says Hackett, "And oh, look, suddenly we've discovered 80% of the kids in this school district don't have an adequate connection." Borgen noted that the digital divide within the county "really hit home during the pandemic" as "some students could not connect … with their teacher because they didn't have broadband and access, didn't have connections, didn't have devices, [and] the families didn't know how to connect." For the COE director and educators in Santa Cruz, action had to be taken to "[make] sure [they] had a mechanism to give students access and … be on the same playing field …[and] have the same opportunities as their more privileged peers" according to Borgen,

> Something had to be done quickly, so the COE and Cruzio came together in March 2020 to imagine ways to provide students with Internet connections. The first idea, executed by April 2020, was extending the Internet connection already available in schools to their parking lots using Cruzio's wireless technology (Dolgenos et al., 2020).

This allowed students and their families to park and access the Internet while maintaining social distancing. However, this was not enough. Students living in rural areas depended on school buses to get to school, but the buses did not run while the schools remained closed, and students therefore could not access free Internet. A director at Community Foundation Santa Cruz County recalls stories of students from remote areas sitting outside fast-food restaurants near Santa Cruz with their laptops open trying to attend school with the restaurants' free Wi-Fi. More had to be done.

Partnership: It Takes a Trio

Equal Access Santa Cruz (EASC) was created out of the necessity to do more to address broadband inequalities in the county. By June 2020, the strategy had shifted to bringing connections to the rural and low-income neighborhoods where students lived. The key to EASC's success lies in the partnership between a local ISP, the school districts, and the Community Foundation. The ISP provided the technology and equipment, the school districts helped identify those in need and provided valuable real estate for network expansion, and the Community Foundation provided the financial resources.

The school districts and COE were important to EASC because they had the most intimate knowledge of those in need of broadband. Hackett noted that working with school districts has been "a real key to how [EASC has] been able to be successful because [the school districts] already had access to all that data [regarding whether families have a connection or not]" as these families are the school district's constituents. The COE created a survey that connected with the individual school districts to verify whether students qualified for affordable broadband service, which saved Cruzio and the county significant administrative costs around eligibility and ensured students' information stayed private.

Further, schools aided the physical expansion of EASC's network. According to Hackett, schools in the county "tend to be distributed evenly around a community and the folks who go to the school are in a pretty nice catchment area around them." Therefore, schools become ideal wireless Internet distribution hubs. Coordination between COE, the school districts, and Cruzio allowed the ISP to access the rooftops and place wireless equipment, helping EASC expand without having to apply for additional permits elsewhere and seek alternative locations.

The Community Foundation also helped the project with financial resources. When Cruzio realized that it needed more funding to expand the network quickly, it sought help from the Foundation. A special fund for the EASC was established at the Foundation. The Foundation believed that instead of relying on small donations from ordinary individuals, it needed "big gifts to power the [project] early" as expanding the network required paying for equipment costs upfront, according to Heuer. Its outreach strategy paid off when Driscoll, a locally based major agricultural company, and other major donors tied to Silicon Valley contributed $500,000 to EASC. The money became the seed grant for a demonstration project in Buena Vista Migrant Center and ensured some families could get up to two years of free broadband and others at a discounted rate (Monroy, 2021). The local Rotary Club and individual community members also chipped in to help connect underserved and unserved communities.

The Foundation also helped coordinate media outreach that further boosted donations to the project. The narrative of helping rural and underprivileged students gain Internet access and facilitate their education really struck a nerve in the broader Santa Cruz community. Heuer said "people that have always been passionate about giving to kids and education just saw this as something they really wanted to be a part of, trying to help close this [broadband] gap." Reflecting on the success of the donation campaign, Heuer continues,

> [Many donors] had kids, they had grandkids, they knew how hard it was for any kid during remote learning. And to know that some kids didn't have a device, were off the map, that opened up a lot of checkbooks. And that gave [EASC] the flexibility to focus on places where the need was greatest, rather than maybe the lowest hanging fruit or what was technically already very easy to do, but just needed money. We really dug in and said, let's try and help those that have the most at stake here of falling behind.

Similary, Hacket reflects that

> The Community Foundation and the philanthropic efforts that they made to fund the whole thing…gave Cruzio the ability to have the seed capital that allowed [it] to build into those areas that would never make sense from a purely business perspective. says Hackett.

Cruzio translated the partnerships and its experience working in the area into timely completion of fast broadband services for students and county residents. When asked about why the company opted for wireless technology as opposed to fiber-to-the-premise (FTTP), Hackett stated that "there's nothing like fixed wireless for speed of deployment and bang for your buck." Using the Terragraph technology developed by Facebook, Cruzio was able to guarantee homes in the service catchment area 100 Mbps symmetrical with the possibility of expanding to full gigabit symmetrical using just wireless technologies according to Hackett. Additionally, Cruzio built its networks with ample redundancy in mind to ensure low latency on its networks. With EASC and the school districts' efforts, the percentage of students without connectivity has decreased from 20% pre-pandemic to around 3%–5%.

An Eye for the Future

While it is common for local programs to be discontinued after initial funding runs out, EASC faces less risk, as program and financial sustainability are built into program design. The combination of new wireless technology that can support many users simultaneously and Cruzio's redundancy-focused network-building approach allowed EASC distribution points to offer service to both subsidized and regular consumers without degrading service quality. Qualified users pay just $15 a month or receive free Internet services and regular customers pay the market price; the service is the same for both sets of customers and there is no distinction between the two groups in Cruzio's network. Having market-rate customers help pay for some of the operating expenses of EASC serves as a "golden handcuff" to Cruzio since "there's no way [it] could close [EASC] down without having a hit on [its] business," Hackett recounts.

With the success of early EASC pilots, Santa Cruz County and nearby Monterey County pledged $500,000 and $350,000 respectively from their ARPA funds for Cruzio to construct additional EASC wireless hubs (Tovar, 2022). The success and speedy implementation of the partnerships with local school districts and the Foundation served as a "shovel-ready" and proven method to expand broadband access, giving Cruzio an easier time with legislators and grant supervisors, as individual donations to the EASC will inevitably dry up and grants and public-sector support will be needed to connect more isolated areas. Hackett said that EASC's earlier success

> makes it much easier for folks in county government to say, 'That's something that we can fund. That's something that makes sense to us.' … Even the most tech-phobic county supervisor can understand [how EASC works and how it will succeed] and get behind it and say okay, that sounds like good value.

The EASC project also has led to broader community synergies to address digital equity issues. The COE and school districts are looking to expand their digital literacy training programs via family outreach and other tactics. The county is now part of a regional digital literacy partnership shiftED, which works with the California Department of Education to build a digital literacy roadmap that includes developing digital lessons for core curriculum, parent training, and evaluations (Santa Cruz County Office of Education, 2022). The COE and the Foundation also have submitted a new grant request to California Public Utilities Commission to address the digital divide by providing connectivity, devices, and digital training for parents.

Mechanisms for Local Action on Broadband

The community resilience framework introduced in this chapter helps analyze the various mechanisms for local action on broadband. Examining the two cases in detail shows that the development and engagement of community resources, the presence of active agents, a project's positive impact, and the opportunities provided by the pandemic served as mechanisms that enhanced broadband delivery.

Development and Engagement of Community Resources

Undoubtedly, having more community resources positively impacts broadband development. Urban and suburban areas in the US have consistently been ahead of rural areas in terms of broadband penetration (Vogels, 2021). Low population density, which results in high cost of broadband deployment, has been cited repeatedly by ISPs as the main reason not to provide service to rural America, a challenge that various state's policies have tried to address, as shown in Chapter 3.

However, these case studies make clear that the initial lack of resources can be overcome. Both locations suffer from low population density, persistent poverty, and rough topography. Neither location had the financial capital outright to expand broadband, nor sufficient political capital to pressure existing providers to expand broadband access. Brownsville's attempt was criticized heavily by incumbent providers, none of which bid for the expansion project; and AT&T withdrew DSL service from Buena Vista Migrant Center in Santa Cruz County. While Santa Cruz County had local ISPs with sufficient technical knowledge to expand broadband access, it lacked the financial capacity to realize such projects by itself. Linkages between the institutions and actors played major roles in pushing forward broadband projects in both locations.

The development and engagement of community resources are one key mechanisms responsible for delivering broadband projects in both locations. In Brownsville, the city's relationship with its eventual private partner was born out of initial engagement meetings. The broadband survey showed that the city's residents overwhelmingly lacked reliable and fast access, and provided the city administration with clear evidence regarding the incumbent providers' opposition. Further, Lit's commitment to training a local broadband workforce will develop the local capacity to maintain and even expand broadband networks, potentially enhancing future broadband delivery.

In Santa Cruz County, EASC's success came from its development and engagement of the ISP, the Foundation, and the COE. Their decision to utilize each other's institutional strengths made the project work. The development of social connections between these institutions was key. In addition, the Foundation was able to utilize the pandemic and the local media to secure donations that provided the seed money to jumpstart the project.

Active Agents

Several key actors played important roles that led to broadband success in both localities. These actors utilized their organizations and expertise to draw attention to the issue, engaged in direct communication with decision-makers, and facilitated project implementation. Active agents in some cases even rallied financial support for the project and helped fend off opposition.

Policy entrepreneurs utilized data and expertise inside the system. Brownsville's broadband success can be traced to earlier research and advocacy work by the former Federal Reserve Bank of Dallas researcher who demonstrated local broadband needs and facilitated the city's eventual success. Similarly, Santa Cruz County's Department of Education staff understood the connection challenges their students faced when shifting to online education, and were able to initiate contact with Cruzio that eventually led to action on broadband.

Active agents utilized their connections with local decision-makers to advance broadband projects, resulting in the partnership between Lit and Brownsville and Cruzio and Santa Cruz education officials. Active agents also use innovation from elsewhere to guide local program development. In Brownsville, the researcher presented best practices for broadband development to city officials and stakeholders. The municipal staff and project partner, supported by local knowledge and data, offered strong and convincing counterarguments against the narrative provided by national ISPs attempting to impede the city's broadband efforts. In Santa Cruz County, Cruzio collaborated with Facebook and utilized new Terragraph technology to deliver fast wireless connections. Further, active institutions and individuals in the county helped rally financial resources that proved crucial to project implementation.

Collective and Strategic Action

Collective action builds from trust and reputation. In Santa Cruz County, both the Foundation and Cruzio have existed in the region for many years and have established themselves as trustworthy local organizations. Heuer said that

> if this were AT&T, [the project] would [not] have worked...Cruzio was a very well-respected local company. There's a lot of very fierce 'buy local' mentality here and support for small businesses. And Cruzio has been around for 30 years. It's a lot of the same people, same family that kind of owns it. And they had a lot of trust to go off.

The strong community reputation of the two organizations helped establish trust in the early stages of the project, leading to a quick and successful fundraising campaign that kickstarted the project.

By contrast, in Brownsville, the eventual project partner, Lit Communities, was a relatively new company in the broadband field. The fact that an executive was born and raised in Brownsville, helped bring Lit to the table in earlier planning phases. However, the city's decision to undergo an open bidding process for both the initial broadband survey and for network construction, shows the city was willing to work with other partners as long as they met certain objective standards set by the city.

In terms of common agenda, Brownsville's agenda of improving broadband access for its citizens aligned well with Lit's mission of engaging the community and providing community-oriented solutions to the digital divide. In Santa

Cruz, the Foundation's mission aligned well with Cruzio's focus on investing in the local community and the county's education mission. These common agendas reduced communication barriers and led to the fast implementation of projects.

Mutually reinforcing activities allow each party to do what it does best. In Santa Cruz, the ISP focused on building the network with their technical expertise, the Foundation concentrated on fundraising which sped up the initial deployment, and the school districts used their connections with the community to ensure qualified households and students received necessary services. The collaboration meant that no agency had to work outside its knowledge base. In Brownsville, the city government facilitated communication with the local public utilities to smooth out potential issues regarding pole access, and the private provider, by taking full responsibility for network management and customer service, helped remediate the city's lack of capacity to operate a commercial broadband network.

Constant communication was key in both cases. In Brownsville, the city organized strategizing sessions with local stakeholders from the very beginning, and was in close contact with the project's bidders throughout the contract negotiation process. In Santa Cruz County, the three major stakeholders all mentioned frequent communication especially during the earlier phase of the project as they deliberated about project details and timeline.

In both cases, one or more of the participating organizations was directly responsible for managing the operations of the projects, serving as a backbone support organization. Both private ISPs and local government units had sufficient planning and management experience thanks to each organization performing mutually reinforcing activities.

Equity and Impact

Equity is present in the program design of both cases. In Brownsville, the broadband project aims to deliver fiber to all homes and businesses inside the city. Lit confirmed that it would participate in the federal ACP program and that it was "committed to working with [its] philanthropic partners and [its] foundation partners and non-profits to essentially, whenever the time comes, if need be, to create an ACP-like program." The ACP exhausted its funding in May 2024 and was not renewed by Congress. It remains to be seen how community partners in Brownsville will secure sufficient funding to ensure a local version of the ACP. Lit Communities is under new ownership, which may not retain the same vision to serve people who were left behind by the big companies.

In Santa Cruz County, Cruzio has built, in its EASC program structure, mechanisms that will continue to guarantee qualifying households pay reduced costs after the end of ACP. However, its network operates on a smaller scale in comparison to the Brownsville project and is not universally accessible to all residents in the county.

Communities need to pursue broader goals – beyond digital equity – to explore how digital inclusion might open up other opportunity structures in health, education, and the economy (Rhinesmith et al., 2023). In Brownsville, the city is using the project to enhance digital literacy and leverage the city's backbone network and wireless technologies to expand broadband access to nearby towns. In Santa Cruz County, Cruzio is already in conversation with other regional ISPs to expand the tri-party model used in the county to nearby access-poor counties and towns.

COVID-19 Pandemic: Crisis and Opportunity

"A crisis is a terrible thing to waste," the economist, Paul Romer, once said (Rosenthal 2009). Of all the mechanisms examined here, arguably none played a greater role than the COVID-19 pandemic in highlighting the need to enhance broadband delivery in the US. The pandemic concretized, in painful ways, the devastating impacts of the digital divide. No longer were the digitally disconnected so invisible (Turner Lee, 2024). The pandemic's exposure of broadband issues persuaded the federal government to allocate substantial new investment via ARPA and the Infrastructure Investment and Jobs Act of 2021 as described in Part 1 of this book.

Both communities seized the opportunity. In Brownsville, while the broadband project started prior to the pandemic, the pandemic significantly accelerated its effort and ARPA funding drastically reduced the city's funding burden. In Santa Cruz County, the pandemic and the subsequent school closure were the impetus of the project and led donors to make substantial donations that facilitated a faster start to EASC.

Moving Forward

Solutions to rural broadband deployment require a multi-level governance approach as described in Chapter 1. This chapter has elaborated a local resilience framework to articulate the local elements that help ensure success in a multi-level governance system. Solutions can be developed at the local level if empowered by coordinated policies at the state and federal level. For federal and state governments, the time has come to pay closer attention to local approaches, to identify the mechanisms of success, and to encourage and support those mechanisms.

Local activists, partnerships, and coordinated action can combine to meaningfully improve local broadband, especially now that new funding is being provided via ARPA and IIJA. The chapters that follow in this section showcase how state policy and local community action helped address broadband deployment in Minnesota (Chapter 5), and Colorado and Maine (Chapter 6). Chapter 7 explores how communities in states that restrict broadband use partnerships to get beyond those restrictions.

References

Ali, C. (2021). *Farm Fresh Broadband: The Politics of Rural Connectivity*. Information Policy Series. Cambridge, MA: The MIT Press. https://doi.org/10.7551/mitpress/12822.001.0001

Ashmore, F. H., Farrington, J. H., & Skerratt, S. (2017). Community-led broadband in rural digital infrastructure development: Implications for resilience. *Journal of Rural Studies*, *54*, 408–425. https://doi.org/10.1016/j.jrurstud.2016.09.004

Ashton, W., & Kelly, W. (2019). Innovation, broadband, and community resilience. In Ed. by Matteo Vittuari, John Devlin, Marco Pagani, and Thomas G. Johnson (eds.), *The Routledge Handbook of Comparative Rural Policy* (pp. 391–408). Routledge. https://doi.org/10.4324/9780429489075

Barton, J. (2021). Planning the infrastructure of the digital economy. *The Architectural League of New York*. Retrieved from https://archleague.org/article/brownsville-digital-economy-infrastructure/

Barton, J., Perlmeter, E. R., & Marquez, R. R. (2015). Las Colonias in the 21st century: Progress along the Texas-Mexico border. *Federal Reserve Bank of Dallas*. Retrieved from https://www.dallasfed.org/~/media/documents/cd/pubs/lascolonias.pdf

Cameron, C. (2022, June 2). Jordana Barton-García becomes Connect Humanity's first fellow. *Connect Humanity*. Retrieved from https://connecthumanity.fund/jordana-barton-garcia-becomes-connect-humanitys-first-fellow%ef%bf%bc/

Casper, J. (2021, September 15). The state of state preemption – Seventeen is the number. *Community Networks*. Retrieved from https://muninetworks.org/content/seventeen-states-preempt-municipal-broadband

City of Brownsville. (2022). Cultural and tourism grant fund. Retrieved from https://www.brownsvilletx.gov/2297/Cultural-and-Tourism-Grant-Fund

DePietro, A. (2021, November 26). U.S. poverty rate by city in 2021. *Forbes*. Retrieved from https://www.forbes.com/sites/andrewdepietro/2021/11/26/us-poverty-rate-by-city-in-2021/

Diaz-Pineda, Z. (2021). Local chronic financial illness. *The Architectural League of New York*. Retrieved from https://archleague.org/article/brownsville-local-chronic-financial-illness/

Diaz Torres, P., & Warner, M. E. (2024). A policy window for equity? The American Rescue Plan and local government response. *Journal of Urban Affairs*. Advance online publication. https://doi.org/10.1080/07352166.2024.2365788

Dolgenos, P., Rodriguez, M., Sabbah, F., & True, S. (2020, October 6). Crucial to provide internet connectivity for students. Santa Cruz Sentinel. Retrieved from https://www.santacruzsentinel.com/2020/10/06/guest-commentary-crucial-to-provide-internet-connectivity-for-students/

Doussard, M., & Schrock, G. (2022). Urban policy entrepreneurship: Activist networks, minimum wage campaigns and municipal action against inequality. *Urban Affairs Review*. https://doi.org/10.1177/10780874221101530

Fawcett, S. B., Paine-Andrews, A., Francisco, V. T., Schultz, J. A., Richter, K. P., Lewis, R. K., Williams, E. L., et al. (1995). Using empowerment theory in collaborative partnerships for community health and development. *American Journal of Community Psychology*, 23(5), 677–697. https://doi.org/10.1007/BF02506987

Flora, C. B., Flora, J., & Gasteyer, S. (2018). *Rural Communities: Legacy and Change* (5th ed.). Routledge. https://doi.org/10.4324/9780429494697

Gillett, S. E., Lehr, W. H., & Osorio, C. (2004). Local government broadband initiatives. *Telecommunications Policy*, 28(7), 537–558. https://doi.org/10.1016/j.telpol.2004.05.001

Gonsalves, S. (2021, July 7). The problem(s) of broadband in America. *Institute for Local Self-Reliance*. Retrieved from https://ilsr.org/wp-content/uploads/2021/07/Problems-of-Broadband-072021.pdf

Grubesic, T. H., & Mack, E. A. (2015). *Broadband Telecommunications and Regional Development*. Routledge Advances in Regional Economics, Science, and Policy. New York: Routledge. https://doi.org/10.4324/9781315794952

Guo, Y. E. (2023). *Mechanisms for Local Broadband Delivery: A Case Study*. Unpublished thesis, Cornell University, Ithaca, NY. https://hdl.handle.net/1813/116892

Hlavaty, C. (2013, October 30). Brownsville named the poorest city in America. *The Houston Chronicle*. Retrieved from https://www.chron.com/news/houston-texas/texas/article/Brownsville-named-the-poorest-city-in-America-4939821.php

Isenberg, S. (2018, May 22). SC County broadband survey shows demand for internet service. *Santa Cruz Tech Beat*. Retrieved from https://www.santacruztechbeat.com/2018/05/22/broadband-survey-shows-demand-for-internet-service/

Kania, J., & Kramer, M. (2011). Collective impact. *Stanford Social Innovation Review*. Retrieved from https://ssir.org/articles/entry/collective_impact

Lit Communities. (2022). BTX fiber. Retrieved from https://litcommunities.net/wp-content/uploads/2022/02/lit-com-case-study-BTXFiber-digital-w-back.pdf

MacWillie, L., Menzel, K., Miller, J., & Ramirez, J. (2021). Brownsville undercurrents. The Architectural League of New York. Retrieved from https://archleague.org/article/brownsville-texas-intro/

Magis, K. (2010). Community resilience: An indicator of social sustainability. *Society & Natural Resources*, 23(5), 401–416. https://doi.org/10.1080/08941920903305674

Mendez, T. (2022, March 10). Bridging the digital divide with American Rescue Plan Act funding. *Route Fifty*. Retrieved from https://www.route-fifty.com/infrastructure/2022/03/arpa-broadband-digital-divide/362954/

Monroy, L. (2021, April 27). How equal access Santa Cruz county is bridging the digital divide. *Good Times*. Retrieved from https://www.goodtimes.sc/equal-access-santa-cruz-county-bridging-digital-divide/

National Digital Inclusion Alliance (NDIA). (2017). Worst connected cities 2017. Retrieved from https://www.digitalinclusion.org/worst-connected-cities-2017/

National Digital Inclusion Alliance (NDIA). (2018). Worst connected cities 2018. Retrieved from https://www.digitalinclusion.org/worst-connected-2018/

National Digital Inclusion Alliance (NDIA). (2019). Worst connected cities 2019. Retrieved from https://www.digitalinclusion.org/worst-connected-cities-2019/

Ostrom, E. (2010). Analyzing collective action. *Agricultural Economics*, 41(s1), 155–166. https://doi.org/10.1111/j.1574-0862.2010.00497.x

Rhinesmith, C., Dagg, P. R., Bauer, J. M., Byrum, G., & Schill, A. (2023). *Digital Opportunities Compass: Metrics to Monitor, Evaluate, and Guide Broadband and Digital Equity Policy.* Quello Center, Michigan State University, Lansing, MI. Retrieved from https://quello.msu.edu/wp-content/uploads/2023/02/Digital-Opportunites-Compass-Paper-20220223.pdf

Roberts, E., Anderson, B. A., Skerratt, S., & Farrington, J. (2017). A review of the rural-digital policy agenda from a community resilience perspective. *Journal of Rural Studies, 54,* 372–385. https://doi.org/10.1016/j.jrurstud.2016.03.001

Rosenthal, J. (2009, July 31). A terrible thing to waste. *The New York Times.* Retrieved from https://www.nytimes.com/2009/08/02/magazine/02FOB-onlanguage-t.html

Santa Cruz County Office of Education. (2022, September 20). ShiftED: COE and CDE launch digital literacy collaborative. Santa Cruz County Office of Education. Retrieved from https://santacruzcoe.org/shifted-coe-and-cde-launch-digital-literacy-collaborative/

Silverman, D. (2019, June 3). Best internet service in Texas? It might be in tiny Mont Belvieu. *Houston Chronicle.* Retrieved from https://www.houstonchronicle.com/techburger/article/Best-internet-service-in-Texas-It-might-be-in-13908274.php

Sylvain, O. (2012). Broadband localism. *Ohio State Law Journal, 73,* 795.

Taylor, S. (2022a, July 5). Charter Communications does not like Brownsville's new universal access broadband network. *Rio Grande Guardian.* Retrieved from https://riograndeguardian.com/charter-communications-does-not-like-brownsvilles-new-universal-access-broadband-network/

Taylor, S. (2022b, July 7). Brownsville mayor dismisses criticism of city's broadband plan by AT&T, Spectrum. *Rio Grande Guardian.* Retrieved from https://riograndeguardian.com/brownsville-mayor-dismisses-criticism-by-att-spectrum-of-citys-broadband-plan/

Tomer, A., Fishbane, L., Siefer, A., & Callahan, B. (2020). Digital prosperity: How broadband can deliver health and equity to all communities. *Brookings.* Retrieved from https://www.brookings.edu/wp-content/uploads/2020/02/20200227_Brookings-Metro_Digital-Prosperity-Report-final.pdf

Tovar, R. (2022, February 1). Santa Cruz County expands broadband program to 1,200 households. *KION546.* Retrieved from https://kion546.com/top-stories/2022/02/01/santa-cruz-county-board-approve-partnership-to-expand-broadband-acess/

Treacy, A. (2022, August 26). Brownsville, Texas and Lit Communities partner to build citywide fiber network. *Community Networks.* Retrieved from https://muninetworks.org/content/brownsville-texas-and-lit-communities-partner-build-citywide-fiber-network/

Turner Lee, N. (2024). *Digitally Invisible: How the Internet Is Creating the New Underclass.* Washington, DC: Brookings.

U.S. Census Bureau. (2021). Presence and types of internet subscriptions in households (2021 ACS 5-year estimates). Retrieved from https://data.census.gov/table?q=B28002&g=0400000US48&tid=ACSDT5Y2021.B28002

Vogels, E. A. (2021, August 19). Some digital divides persist between rural, urban and suburban America. *Pew Research Center.* Retrieved from https://www.pewresearch.org/fact-tank/2021/08/19/some-digital-divides-persist-between-rural-urban-and-suburban-america/

5

MINNESOTA – AN EARLY LEADER IN ADDRESSING RURAL BROADBAND

Y. Edward Guo, Elizabeth H. Redmond, and Mildred E. Warner

* * *

St. Louis County, Minnesota stretches from the Canadian border to the western tip of Lake Superior, bifurcated by a belt of strip mines. The county is sparsely populated and deeply tied to the iron industry, producing the majority of the United States' domestic supply. Located just east of the largest open-pit iron mine in the state sits Cherry Township which, while home to fewer than 1,000 residents, is an exemplar in community-led broadband advocacy. In 2018, St. Louis County had lower broadband penetration rates than most of its peers, and the vast majority of Cherry residents – 446 households – had no broadband service at all, while others – 8 households – had less than 10/1 Mbps (Kruse, 2018a). Professionals could not work from home, farmers couldn't sell their crops, and elderly residents couldn't connect their medical devices to critical services. There were stories of young adults who refused to come home for Christmas, and towns that couldn't replace their sole doctor because no one would move somewhere without Internet connectivity. Faced with limited public resources and little private incentive, Kip Borbiconi, a long-time resident and retired mine worker, took on the herculean task of bringing broadband to Cherry.

Borbiconi's foray into broadband advocacy was more accidental than intentional. In the mid-2010s, Cherry was visited by a state representative who was seeking local funds to organize what later became the Iron Range Broadband Committee, which brought together local communities to collectively seek out broadband grants. Cherry donated to the cause and joined forces with nearby Chisholm/Balkan Township, Hibbin and Mt. Iron/Buhl (Kruse 2018b). Borbiconi, a lifelong technology enthusiast and frequent attendee of town board meetings, became the de facto committee liaison.

DOI: 10.4324/9781003619208-7

Understanding that their individual towns were too small to build or operate any potential broadband network, the group met with various providers to serve the larger region. In 2018, they hired an outside consultant to help draw up a plan toward accessibility (Kruse, 2018b). The committee agreed to "go big or go home," and opted to pursue a future-proof fiber buildout. After studying various funding mechanisms, the group eventually applied for a USDA ReConnect grant. Despite their original goal of regional service, the stringent grant criteria required the application be limited to Cherry.

With a large grant application in his back pocket, Borbiconi began going door to door in Cherry to gather support – something he had never done before. To increase the application's chance of success, Borbiconi needed to identify 20 businesses in the proposed service area; he also needed to demonstrate that there was requisite demand from the community as a whole. Even Borbiconi, who relied solely on his cell phone's hotspot to connect to the Internet, was surprised by the overwhelming accessibility needs of his community.

Borbiconi found a willing partner in CTC – a Minnesota-based telephone, cable, and Internet cooperative. CTC was one of the few telecommunication providers in the area that came to the negotiation table while the committee was trying to identify a private partner for network construction and operation. When CTC representatives came to the community for a preliminary assessment, they were blown away by the turnout, noting that they had "never seen a town hall so packed." Borbiconi's persistent actions, from canvassing and organizing to handing out fliers, had united the town. He commented that the cookies and coffee he had brought were "nowhere near enough" for the crowd that had come to support the grant and the CTC partnership.

* * *

Minnesota has been lauded for its approach to expand broadband access, boasting one of the oldest programs in the nation (The Pew Charitable Trusts, 2020b). The state aims to have all businesses and homes connected by one provider offering Internet service of at least 100/20 Mbps by 2026 (Minnesota Department of Employment and Economic Development, 2024). To do this, the state has implemented successful programs that target rural communities, bring local advocates to the table, and enhance the visibility of non-traditional providers. Minnesota's broadband policies focus on providing place-based subsidies to encourage private market actors to expand broadband coverage. Consequently, local actors utilizing state broadband programs have become a major vector through which projects are being realized in Minnesota. These local actors facilitate early project planning as well as strengthen channels of communication and data sharing between advocates, government agencies, and service providers. Additionally, we see partnerships among rural Minnesota communities which understand that a regional mindset provides a path toward vital state funding.

However, rural communities in Minnesota have faced the same obstacles which hinder broadband expansion across America. These issues are complex and span the geographic, economic, and political realms. Rural communities are limited by their physical geography, which in this case is difficult and diverse, dotted with lakes, forests, mountain ranges, and mines. Low population density further prevents broadband development. Communities which do not offer vital Internet connection are effectively barred from economic development and this intensifies the rural brain drain. Rural communities face incumbent providers that refuse to update old infrastructure, and offer inadequate service. Lack of revenue to support expansion forces communities to be dependent on grant funding. There are also deeply entrenched political impediments to rural broadband expansion, including overlapping state and federal authority, restrictions on local government, and regulatory barriers.

Minnesota's Broadband Programs

Established in 2014, Minnesota's capstone Border-to-Border Development Grant Program is one of the oldest broadband programs in the US, providing funding for middle- and last-mile infrastructure in unserved and underserved areas (The Pew Charitable Trusts, 2020a). Having been in operation for almost a decade, Minnesota's efforts to expand rural connectivity have supported regional providers, addressed mapping inaccuracies, and promoted collaboration with local community organizations. The program's efforts are complemented by the Blandin Foundation, a non-profit that assists rural communities with grants for planning and adoption initiatives and helps them navigate the application process for state broadband funding (The Pew Charitable Trusts, 2020a). Since its inception, the Border-to-Border program has awarded over $290 million in grant money across the state (Minnesota Office of Broadband Development, 2024).

The Border-to-Border program requires that grantees match 50% of project costs. However, to ensure funds reach low-density areas which cannot meet the requirement, the state also launched a dedicated program that allows up to 75% of the project cost to be matched by the state (Benton Institute, 2022). Minnesota also provides funding for line extensions: individual residences submit applications, and the state reaches out to local ISPs to serve the locations (Fischer, 2023).

Minnesota's policies both expand and restrict the role of local actors. On the one hand, the state supports rural cooperatives by matching local funds and/ or complementing state funding with additional grants and loans. The state also promotes local involvement by requiring that applicants provide proof of community support. On the other hand, local governments are restricted from owning and operating their own broadband networks unless they obtain an oner-ous referendum supported by 65% of voters (Cooper, 2023).

Additionally, because federal agencies and broadband offices do not neces-sarily coordinate, states are not always informed regarding which areas have received federal funding (Institute of Local Self Reliance, 2016). To avoid dupli-cative funding, Minnesota allows ISPs to challenge applications involving areas where these ISPs already provide service or where construction is underway. ISPs must contribute to the development of state maps to submit a challenge (The Pew Charitable Trusts, 2020a). Still, challenge processes have been the subject of criticism for protecting private investment over the public good, and failing to require that incumbents upgrade their infrastructure (Institute of Lo-cal Self Reliance, 2016). Challenge processes can overwhelmingly burden un-derserved communities with access to slow, unreliable, and expensive Internet service.

The visibility and complexity of Minnesota's state-level broadband policy was quite unique at the time of Border-to-Border's enactment. Prior broadband policy was often dictated at the federal level and implemented at the local level with little state involvement. However, with Minnesota leading the pack, it has become clear that state policy can play a large role in translating federal-level policy to local-level outcomes. It is important that we study the policy dynam-ics in this multi-level governmental landscape. We ask: (1) How important is the role of the local advocate in rural broadband efforts? How do they promote community resilience? (2) In what ways do non-traditional providers, especially at the local and regional level, aid in expanding broadband access? How do they interact with local, state, and federal-level policy? (3) What barriers exist for rural communities and what is the impact of state policy in removing or uphold-ing them?

The analysis in this chapter is based on expert interviews with key actors in several state-funded broadband initiatives we conducted in 2023 and 2024 (Bravo and Warner, 2024; Guo, 2023). The cases were selected due to their ru-ral nature and the involvement of non-traditional providers. Recommendations from Minnesota's Office of Broadband Development were key to the identifica-tion of key players from local government, local ISPs, and local advocates.

Community Resilience and Active Agents

As discussed in Chapter 4, broadband infrastructure development is an important component of community resilience. Broadband development requires the coop-eration of both the public and private sectors, and benefits from active agents operating across jurisdictional levels. In rural areas, active agents at the com-munity level are crucial to the success of broadband programs, as they are able to identify weaknesses from the bottom up. Additionally, these community-level active agents often act as a proxy for their local government in broadband efforts, as small townships and villages seldom have dedicated broadband offices or officials. It is not uncommon that volunteer task forces communicate with ISPs and apply for state funding, acting for their local government. Importantly, state policy acts as the bridge between these agents' efforts and top-down federal policy and funding. These active agents leverage existing community resources to support broadband expansion (Figure 5.1).

The broadband effort in Cherry was so successful due to the cooperation of active agents across the state, at the regional, local, and community levels. Each actor sought to leverage existing local resources to develop critical broadband infrastructure: the state representative aimed to tap into local funding to bring the Iron Range Broadband Committee to life; the regional coalition aimed to agglomerate local population and demand to appeal to ISPs; the local government relied on a community member to act as their liaison; and Borbiconi applied his lifetime of experience as a Cherry resident to bring the community together and prove to the ISP that the town was worth serving.

It is important to note that at each jurisdictional level, active agents were looking to leverage local resources. Given that proof of community support is required for Minnesota's Border-to-Border grant applications, local resources, both monetary and human, are essential for success; entirely top-down or private sector applications are likely to fail. This state requirement gives local governments and local advocates a large role in planning for their own communities; as

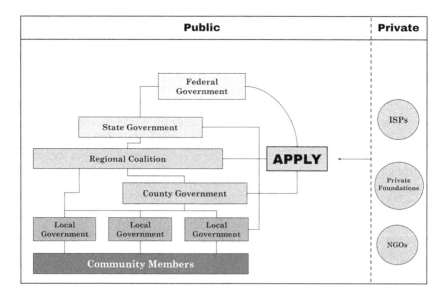

FIGURE 5.1   Network of actors in the multi-level governance framework. Figure by authors.

Borbiconi notes, "our politicians can go so far, but it's the common people that make it happen."

Given that St. Louis Co. has historically lacked the resources to fund local projects, the type of community-driven success shown in Cherry has become the new model. As new federal funding streams have been established in the wake of the COVID-19 pandemic, the county has begun actively supporting community-led efforts. Beyond dedicating $2 million of its $54 million ARPA funding to further such efforts, the county and regional economic development organizations have begun providing additional technical and financial support (St. Louis County, 2021a; 2021b).

Le Sueur County

St. Louis County is not alone in this community-driven approach to broadband expansion. Le Sueur Co. is a largely rural county located outside the twin cities of Minneapolis and St. Paul, Minnesota. With a voluntary task force of residents, the county managed to secure $3 million via three rounds of state funding that, combined with equity from ISP BevComm and other local resources, connected over 800 homes, farms, businesses, and anchor institutions. To make broadband expansion more accessible across the county, Le Sueur Co. provided interest-free loans to certain townships which allowed them to meet their immediate state match requirement.

In Le Sueur Co., local broadband advocates and their voluntary task force worked with the county to initiate contact with local ISPs to understand what each provider was doing in the county and why providers were not building out to unserved areas. These initial contacts opened up lines of communication between the county and providers, like BevComm, who were able to educate advocates on the cost of broadband expansion and the process of a successful state grant application. The voluntary task force also organized a line-extension mail campaign in an attempt to connect unserved homes left behind by incumbent providers. In addition, they allocated COVID-19 relief funding to enhance fiber backhaul and attended county fairs to raise public awareness.

The strategic, collective, and importantly *voluntary* efforts of community members in both St. Louis and Le Sueur Co. were critical to a successful broadband rollout. With so much pressure on residents to support complex infrastructure projects, volunteer task forces can be fragile. Often, the brunt of responsibility falls on the shoulders of a select few, who accept an inordinate amount of work for no pay. As noted by Barbara Kline, a local broadband advocate in Le Sueur Co.:

One thing that…has really been a loss, especially right now, is there was a rural broadband coalition which had been very aggressive with the legislature…

The person who was running that got sick and it just folded. So, we don't have any rural coalition anymore that is taking this on with the legislature.

Without adequate funding, rural municipalities cannot create dedicated government positions to tackle the digital divide. Likewise, a lack of expertise within existing governmental bodies leads to an overreliance on constituents to solve problems which are the government's responsibility to address.

Non-traditional Providers and State Policy

Broadband accessibility is a first-order problem in rural America; without physical infrastructure in place, efforts to increase affordability and adoption are largely moot. State policy in Minnesota disincentivizes municipally owned or operated broadband networks, thereby requiring private provider participation. Even so, state policy revolves around smaller, local, and non-traditional providers like electric and telephone cooperatives (The Pew Charitable Trusts, 2021). As noted by Bree Maki, the Director of Minnesota's Office of Broadband Development, the state prides itself on its strong cooperative model and believes such models enhance accountability: "These [co-ops include] folks that live and work in these communities. They're accountable to each other and just truly are asking 'what do our members need and how can we best serve them?'"

Minnesota policy requires proof of community support, which pushes ISPs to interact with community-level actors to receive grant funding. Non-traditional providers, who operate locally and oftentimes have ties to the towns they hope to serve, make it a point to reach out to local broadband advocates. Likewise, non-traditional providers work closely with local governments throughout the entire application process. Three non-traditional exemplars currently operating in the state include Paul Bunyan Communications, Harmony Telephone (Mi-Broadband), and the Northeast Service Cooperative.

Paul Bunyan Communications Telephone Cooperative

Paul Bunyan Communications is an example of a cooperative ISP which works in partnership with local communities to bring funding to rural areas. Beginning as a telephone company in 1950, it began expanding its service area and offering Internet services after the passage of the Telecommunications Act of 1996. With the goal of bringing economic vitality to the region, Paul Bunyan Communications has been awarded over $15 million in state broadband grants over seven rounds, and has successfully connected thousands of households and businesses to fiber across Central and Northern Minnesota (Minnesota Office of Broadband Development, n.d.). Paul Bunyan's rural work stands out largely because the ISP understands how prohibitive the cost of broadband infrastructure is to local communities. One of the cooperative's most significant projects has been expanding fiber across Itasca, St. Louis, and Hubbard Counties. The project cost $1.78 million; Paul Bunyan contributed $980,990, and the State of Minnesota contributed $802,620 (Paul Bunyan Communications, n.d.).

Paul Bunyan seeks out local broadband advocates to champion their work. In Itasca, St. Louis, and Hubbard, Steve Howard, the Development Manager of Paul Bunyan Communications, spoke highly of the broadband advocates' work in project planning and initiation. The cooperative found that these local-level active agents help successfully distribute surveys that demonstrate community demand, a vital method for gauging possible take rates. Howard noted, "somebody that has a reputation, that hangs out in the coffee shop in the morning or at the bar at night...brings up the topic and makes things happen;" this really piques the interest of providers who understand the need to disseminate information to the people they aim to serve.

MiBroadband Cooperative

Harmony Telephone, now MiBroadband, is a part of a partnership between three area cooperatives: two broadband and an electric cooperative. The collaboration was formed in 2018 to fill the connectivity gap in Southeastern Minnesota and Northeastern Iowa at the regional level. The broadband issue was a key driving force for the partnership as, in the rural counties these cooperatives serve, there are not enough resources to implement these projects. The cooperatives have received over $5 million in state broadband funding for three projects that will ultimately extend fiber to over 500 homes and businesses in rural Fillmore County. The collaboration between the three is the key to their success.

Northeast Service Cooperative

The Northeast Service Cooperative (NESC) is an example of an ISP partnering with last-mile providers and local governments to promote successful broadband expansion. NESC was established by the Minnesota legislature in 1976 as a non-profit public corporation. Serving seven counties in northeastern Minnesota, NESC provides an array of niche services ranging from group health insurance to academic programs.

NESC entered the middle-mile broadband arena in 2011 with the help of $45.3 million from a 50%-grant-50%-loan program federally funded through the USDA Rural Utilities Service. Because NESC focuses on middle-mile fiber, they seek out other providers who specialize in last mile; such partnerships facilitate rural broadband deployment. Paul Brinkman, the Executive Director of NESC, underscores the importance of establishing clearly divided roles in these partnerships and explains that they are not interested in internal competition: "We were able to communicate at the front end...to the providers of the last-mile level that we had no interest in eating their lunch. We wanted to be their part-ner." NESC has doubled its middle-mile fiber network to nearly 1,200 miles since 2014, becoming an important network backbone in the region.

Beyond formal partnerships with last-mile providers, NESC also benefits from its informal relationship with Cooperative Network Services (CNS), a broadband engineering company. When someone tries to connect to a network asset, CNS might call NESC to see how close the NESC's fiber is to the in-quirer's desired location and whether NESC's fiber can be extended to meet potential customer needs. These "structured or even loose network structures, partnerships, and loose network of colleagues allow for opportunities to meet unserved and underserved areas simply because of the collegial relationships you have in the marketplace," says Brinkman (NESC).

Minnesota state policy, in supporting small, non-traditional providers, fosters an atmosphere of cooperation instead of competition, which focuses the atten-tion on the needs of the community instead of the company's market share or stock price.

Barriers for Rural Communities and the Impact of State Policy

While rural communities across America face intrinsic challenges to broadband investment like limited market demand and local fiscal capacity, there are unnec-essary barriers erected at the state and federal level which unduly impact acces-sibility. Most rural communities are dependent on grants to fund their broadband efforts, and those communities depend on the efficacy of state and federal policy design. Any change in grant programs will impact local broadband projects. One of the most prevalent barriers facing rural communities is regulatory mismatch between state and federal programs. In addition, high local match requirements create high barriers to entry that many rural communities cannot reach.

Regulatory Mismatch

Minnesota does not allow state funds to be used in areas where ISPs have been awarded federal funds (Orenstein, 2021); state grant applications are required to remove those areas from their application. One major federal program which often complicates state grant applications is the FCC's Rural Digital Opportu-nity Fund (RDOF). Many ISPs have been granted RDOF funds with no ability to serve the whole awarded area. However, if areas covered by RDOF fund-ing apply for state funding, they are denied. Brad Gustafson, the Community Development Manager in St. Louis County, noted that one provider had been awarded a significant amount of RDOF funds to connect rural communities, but there was "no way that [the grantee was] going to be able to do [those] kinds of projects" at such a large scale. Some communities within the grant area that had already been funded through the state's Border-to-Border program had their grants pulled (Colburn, 2021). Additional issues with the RDOF grant's bidding process led the FCC to temporarily withhold funding, leaving communities in need without the support they were promised (Goovaerts, 2021).

Regulatory mismatch related to the federal RDOF program is widespread. According to BevComm's CEO Bill Eckles, state grant applications need to map out areas that have been pre-awarded federal dollars so as not to include them in the service area. This creates significant roadblocks for broadband project planning and implementation. Le Sueur County discovered how cumbersome the rule was when, in 2020, the FCC announced that $1.3 billion in RDOF funding was awarded to LTD, $311 million of which was going to build out fiber optic cable in Minnesota (Goovaerts, 2022). Rural areas were split by census block, so that only some qualified as "unserved" and would be pre-awarded RDOF funds. When Le Sueur and BevComm applied for a state grant, they knew they would need to avoid including any of the pre-awarded areas; this was practically impossible. Their application was denied.

Eckles noted that this overlap with federal funding areas cost the county a vital project:

> When you looked at Le Sueur County, it looked like a checkerboard. One block would get funded; the one after wasn't…We essentially had to include these areas that had RDOF funding because there was no way to do it. You couldn't zero down into it by census block.

Local communities and providers, doubtful of LTD's ability to serve the claimed areas, quickly and efficiently protested. The FCC revoked the RDOF award in 2022, concluding that LTD was not capable of building the fiber-optic cable networks it had promised (Orenstein, 2023). However, because LTD appealed the decision, it remained unclear whether local service providers could obtain state and other federal funding in those contested areas.

The debate finally appeared settled when the state's Public Utilities Commission (PUC) suspended LTD's certification in late 2023. The time it took to sort out the regulatory mismatch delayed potential shovel-ready projects in rural Minnesota. Paul Bunyan Communications pulled five of six projects due to similar overlapping concerns. Rising costs post-pandemic have made those projects less financially viable than they would have been prior to LTD's award.

Regulatory mismatch also creates difficulties for local broadband projects and providers with limited capacity. One of the biggest challenges faced by rural providers is that the various state and federal program timelines are not aligned. Historic preservation requirements along well-traveled corridors also generate costly planning and project delays. According to Jill Huffman, the COO of Harmony Telephone, different rules across funding opportunities have been difficult to navigate, and meeting the different grant requirements of each funding source can require more effort than is feasible for small providers.

Yet another barrier created by regulatory mismatch centers around broadband mapping. Broadband availability maps are used to inform agencies which areas are eligible for funding; they are also used by ISPs to challenge grant applications submitted by competitors. The disparate policies in place at the state and federal level – even between agencies – create an incomplete picture of broadband availability. To illustrate, Huffman explained that privately funded projects are not recorded anywhere until the next mapping cycle, whereas a USDA-funded project appears on maps as soon as the project is awarded. There is also confusion about how to reflect projects under construction on maps.

Rural communities also struggle to mitigate program limitations, especially as related to challenge processes associated with state grants. To avoid overbuilding in areas with sufficient existing service, Minnesota allows incumbent providers with ongoing plans to serve a given location to challenge grant applications submitted by competitors. While the state makes the ultimate decision about the validity of a challenge, an incumbent provider's opposition can seriously impact the success of grant applications in contested areas.

Incumbent ISPs can claim that they are or will be serving an area within the timeframe of the competing grant application. If the provider defaults on its promise to build in those areas, there are no effective mechanisms at any level to ensure accountability. While the state does prevent defaulting ISPs from participating in future challenge processes, Eckles explained how this did not allay concerns:

> If a provider does challenge and [says] 'No, I'm going to be providing service to this area without grant funding,' there isn't any sort of mechanism to ensure that they ever do. So, in theory, a provider can say, 'No, I'm going to do it just to prevent a competitor from coming into that area, and never actually build it.' And there's no penalties for it.

Neither local communities nor the state's broadband office has the capacity to monitor service providers to ensure projects are completed. In Le Sueur County, addressing these issues with providers will require coordination between federal agencies, the state's Attorney General, and the state's broadband office.

Local Match Requirements

In St. Louis County, the Community Development Manager Gustafson explained that the state's Border-to-Border grant's "50% [maximum state] match [requirement]...is not feasible in [the county]" as local government and smaller

ISPs do not have the necessary funding. It is in the state's interest to award funds to projects that are willing to provide a larger share of local match, as this allows the state to spend its money more efficiently and on a larger portfolio of projects. However, rural and low-income areas – those with the greatest demonstrated need – are often the ones with the hardest time paying for them.

Future program design must consider whether state funds are better spent on awarding projects that require a higher match. One recent pilot program in Minnesota allows for up to 75% state match up to $10 million. This could be significant for smaller, rural communities (Office of Broadband Development Minnesota, 2023). Prior to the pilot program, counties were responsible for mitigating local cost burdens. For example, Le Sueur Co. has provided some townships with interest-free loans to meet their immediate match require-ment for the state program. This loan structure also protects localities from ever-increasing inflation risks. In Fillmore Co., the county provided Harmony Telephone the option for a small loan with 0% interest, to be repaid over three years to help with upfront marketing to ensure a high enough take rate to reach profitability.

Creative Problem Solving

Faced with numerous regulatory and capacity issues, local leaders in Minnesota rely on collective action to come up with novel solutions to their problems. In-spired by the success in Cherry, St. Louis Co. made the innovative decision to emphasize more than just project construction – they supported project planning by dedicating $250,000 of the county's ARPA funding to assist local planning ef-forts (St. Louis County, 2021a). This type of planning grant afforded the county's communities the opportunity to assess demand and identify partners – a critical first step toward a realized project.

Gustafson, the Community Development Manager, noted that the county is "so rural that a lot of the communities…need to do a feasibility study to get a service provider on board to even think about doing a project in their commu-nity." The planning grant, with each grantee eligible to receive up to $25,000 for a 1:1 match, enables municipalities to pay for feasibility studies and, according to Gustafson, help "put a project together, figure out what the rough estimated cost is going to be, and who is going to want the service in the community." With this information, communities can find an ISP willing to construct a broadband network and apply for additional local, state, and federal funding.

Even if projects are not immediately realized, the data gathered and plans developed can still inform and be modified in the future. St. Louis Co. separated the planning grant from their broadband construction grants and this affords mu-nicipal governments more flexibility, as they do not need to immediately propose a shovel-ready project but can instead take a more measured and thoughtful approach.

Counties also support local communities through technical assistance. St. Louis Co. now administers a broadband survey which communities can use to assess demand. The county's broadband survey was originally launched to locate broadband gaps. However, according to Gustafson, the county has gradually found more "success' [administering] it, community-by-community, when they're looking to start the [planning] process." He noted that, "instead of paying a consultant to do a feasibility study, communities will use our survey." The county receives and processes the surveys and returns the findings to communities. In effect, the surveys serve a dual purpose: they generate feedback about broadband deficiencies, and they are able to provide communities an inexpensive way to measure demand.

Technical assistance is also provided by state and regional organizations, including the Blandin Foundation – which has a long and successful track record in Minnesota – and the Iron Range Resources and Rehabilitation Board (IRRRB). Active agents who work across the state, regional, and local levels are critical in building knowledge, contributing to long-term planning sustainability. Whitney Ridlon, a Community Development Representative at IRRRB, expressed how proud she was of all she had learned and how she was now able to offer technical support to communities that will help them make informed decisions about broadband issues:

> [Through planning studies and community engagement], we've gotten to a point now where we have…enough of an understanding where we can go… this is what broadband is, this is what the state goal is, this is why we're here—we're trying to meet the state goal. These are the different providers working in your region or in your township, or next to your township. We'll talk about the maps: here's where you're served and here's where you're unserved. We have a local speed test. Here's the dots in your township and the speeds they're getting.

This type of knowledge creation allows leaders at the state and regional level to offer services to local communities that would otherwise require additional funding. The IRRRB – the state agency which aims to foster economic development in the Iron Range – can now provide services that would otherwise require a community to hire a private consultant as Ridlon remarks:

> In the past, a private consultant has come in and led that and initially, that's what happened. Now, we've found that me and [Gustafson] could go out to a township and deliver that for them at no cost. Then we can connect them with potential providers.

This level of interaction between state, regional, and local entities also means that ISPs have less opportunity to dictate which technology is constructed, as Ridlon continues:

> I've had townships say 'we're okay with fixed wireless. We understand the difference between fiber and DSL, and we're okay that fixed wireless is going to be the solution for our community.' But at least they're understanding it and making the decision on their own versus just Frontier essentially coming in and saying this is just what it is.

Poor policy design at the state and federal level erects barriers to local broadband planning; remedying this issue would require a coordinated and concerted effort on the behalf of all parties – across the public and private sectors. Short of such an effort, circumventing these barriers comes down to active agents engaging in collective action to create novel solutions to each problem. The more state entities, like the IRRRB, engage with local-level projects and the non-traditional providers who implement them, the stronger an asset to communities they become. By creating knowledge at the state and regional level, services which require high-level coordination – like broadband mapping – can be carried out by a public entity at a lower cost to the community than if they were forced to partner with a private corporation.

Minnesota Policy Activates Local Actors and Non-traditional Providers

State governments play a pivotal role in closing the digital divide; they act as the bridge between broad, top-down federal policy and the local communities those policies attempt to aid. Minnesota's approach to broadband has produced marked success by highlighting the role of local actors and non-traditional providers. By requiring community support for state grant applications, Minnesota has afforded local governments and their constituents a seat at the table. Likewise, state agencies like the IRRRB, have begun offering low-to-no-cost essential services to communities which push broadband programs along.

However, as highlighted by Borbiconi's tireless efforts in Cherry, it is often broadband advocates – dedicated volunteers – who work to connect their rural communities. These local advocates feel the impacts of the digital divide firsthand and understand the particular context in which they will have to work. The successes illustrated in Minnesota make it clear that community-level active agents and local governments are the vectors through which rural broadband projects are realized. Supporting these actors is essential for long-term project sustainability. To do this, it is imperative that grant requirements are simplified and made less burdensome. The goals, timelines, and criteria of the different available grant programs need to be better aligned.

Despite generous policy support in Minnesota, there are some areas which remain difficult to serve due to their geography and the rising cost of materials and labor. Greater financial and technical resources will be required in these communities. As one-time funding mechanisms like ARPA begin to dry up, counties, which have been able to subsidize local projects, will no longer be able to help. We must connect the under- and unserved areas as quickly and equitably as possible.

Minnesota's broadband policies shine a light on communities in a way that truly supports broadband accessibility. While there are undoubtedly kinks to work out, especially as related to regulatory mismatch, Minnesota has created a playbook that other states have followed in their state policy design. Rural broadband is a complex task, complicated by a lack of market interest. Thus, rural communities are especially reliant on state and federal policy support. Local involvement is key and should be a fundamental component of policy design.

References

Benton Institute. (2022, December 12). Minnesota again taps Capital Projects Fund to bridge broadband deployment gap. Retrieved from https://www.benton.org/blog/minnesota-again-taps-capital-projects-fund-bridge-broadband-deployment-gap

Bravo, N., & Warner, M. E. (2024). Innovative state strategies for rural broadband: Case studies from Colorado, Minnesota, and Maine. Department of City and Regional Planning, Cornell University. Retrieved from https://labs.aap.cornell.edu/node/882

Colburn, D. (2021, February 3). Border-to-border grant aids Cook broadband: Federal program creates uncertainties for rural broadband development. *The Timberjay*. Retrieved from https://www.timberjay.com/stories/border-to-border-grant-aids-cook-broadband,17279

Cooper, T. (2023, November 17). Municipal broadband 2023: 16 states still restrict community broadband. *BroadbandNow*. Retrieved from https://broadbandnow.com/report/municipal-broadband-roadblocks

Fischer, K. (2023, June 26). States reach the unreachable with broadband line extension programs. *FierceTelecom*. Retrieved from https://www.fiercetelecom.com/broadband/states-reach-unreachable-broadband-line-extension-programs

Goovaerts, D. (2021, August 4). RDOF winners already defaulted on $78M in bids — much more could follow. *FierceTelecom*. Retrieved from https://www.fiercetelecom.com/regulatory/rdof-winners-already-defaulted-78m-bids-much-more-could-follow

Goovaerts, D. (2022, May 17). Telecom group urges Minnesota to revoke LTD Broadband's RDOF cert. *FierceTelecom*. Retrieved from https://www.fiercetelecom.com/telecom/telecom-group-urges-minnesota-revoke-ltd-broadbands-rdof-cert

Guo, Y. E. (2023). *Mechanisms for local broadband delivery: A case study* (Unpublished thesis). Cornell University, Ithaca, NY. https://hdl.handle.net/1813/116892

Institute for Local Self-Reliance. (2016, June 2). Minnesota broadband grant program gets funded, issues remain. Retrieved from https://ilsr.org/minnesota-broadband-grant-program-gets-funded-issues-remain/

Kruse, D. (2018a). Iron Range communities, broadband, executive summary, Cherry Township. Retrieved from https://broadband.ramsmn.org/wp-content/uploads/2019/08/Iron-Range-Communities-Broadband-Executive-Summary-Cherry.pdf

Kruse, D. (2018b). Iron Range communities, broadband roadmap. *NEO Connect*. Retrieved from https://broadband.ramsmn.org/wp-content/uploads/2019/08/Iron-Range-Communities-Broadband-Roadmap.pdf

Minnesota Department of Employment and Economic Development. (2024). Minnesota broadband goals. Retrieved January 17, 2024, from https://mn.gov/deed/programs-services/broadband/goals/

Minnesota Office of Broadband Development. (2024, January 15). Office of Broadband Development 2023 annual report. Retrieved from https://www.lrl.mn.gov/docs/2024/mandated/240085.pdf

Minnesota Office of Broadband Development. (n.d.). Broadband grant program. Retrieved from https://mn.gov/deed/programs-services/broadband/grant-program/

Orenstein, W. (2021, January 29). Why federal grants may set rural broadband in some areas of Minnesota back for years. *MinnPost*. Retrieved from https://www.minnpost.com/greater-minnesota/2021/01/why-federal-grants-may-set-rural-broadband-in-some-areas-of-minnesota-back-for-years/

Orenstein, W. (2023, November 23). Minnesota regulators suspend crucial designation for controversial LTD Broadband. *Star Tribune*. Retrieved from https://www.startribune.com/minnesota-regulators-puc-suspend-crucial-designation-for-controversial-ltd-broadband/600320255/

Office of Broadband Development Minnesota. (2023). Border-to-border broadband infrastructure grant application instructions. Retrieved from https://mn.gov/deed/assets/2024-broadband-grant-instructions-application-guide_tcm1045-613338.docx

Paul Bunyan Communications. (n.d.). Paul Bunyan Communications awarded State of Minnesota Border-to-Border Broadband Grant for portions of Itasca County, St. Louis County, and Hubbard County. Retrieved from https://paulbunyan.net/paul-bunyan-communications-awarded-state-of-minnesota-border-to-border-broadband-grant-for-portions-of-itasca-county-st-louis-county-and-hubbard-county/

St. Louis County. (2021a). St. Louis County broadband planning grant program guidelines. Retrieved from https://www.stlouiscountymn.gov/LinkClick.aspx?fileticket=vQuLSskH_ck%3d&portalid=0

St. Louis County. (2021b). St. Louis County broadband infrastructure grant program guidelines. Retrieved from https://www.stlouiscountymn.gov/LinkClick.aspx?fileticket=98-lr-GM6og%3d&portalid=0

The Pew Charitable Trusts. (2020a). How states are expanding broadband access. Retrieved from https://www.pewtrusts.org/-/media/assets/2020/03/broadband_report0320_final.pdf

The Pew Charitable Trusts. (2020b). Key elements of state broadband programs. Retrieved from https://www.pewtrusts.org/en/research-and-analysis/fact-sheets/2020/05/key-elements-of-state-broadband-programs

The Pew Charitable Trusts. (2021). What policymakers can learn from the "Minnesota model" of broadband expansion. Retrieved from https://www.pewtrusts.org/en/research-and-analysis/articles/2021/03/02/what-policymakers-can-learn-from-the-minnesota-model-of-broadband-expansion

6

COLORADO AND MAINE – REGIONAL AND MIDDLE MILE APPROACHES

*Natassia A. Bravo, Duxixi (Ada) Shen,
Elizabeth H. Redmond, and Mildred E. Warner*

Closing the infrastructure gap in rural areas is a collective effort. In regions where deployment is costly, market size is modest, and resources are scarce, rural communities are partnering with their neighbors and with local providers to solve their connectivity issues. State and federal broadband funding plays a critical role in supporting these partnerships, by offsetting the cost of expanding infrastructure in rural areas. Yet funding is only one component of infrastructure planning. We interviewed individuals in charge of six fiber broadband projects in Colorado and Maine, including representatives from broadband coalitions, local governments, and the broadband industry (for more detail, see Bravo, 2025; Bravo and Warner, 2024). We provide insights on the opportunities and challenges in procuring infrastructure financing and planning. Colorado and Maine are largely rural states that fund both traditional and non-traditional providers, and this chapter shows a varied list of approaches that involve large telecommunications providers, rural electric utilities, and publicly operated broadband networks. Our interviews illustrate the importance of our community resilience framework (described in Chapter 4), as local leaders are innovative within a broader multi-level governance framework (described in Chapter 1).

State Policy Differences in Colorado and Maine

Colorado

Unlike their underserved urban counterparts, rural communities in large states like Colorado face an additional challenge when it comes to infrastructure – the lack of access to a robust middle-mile network. The "middle mile" is a network

DOI: 10.4324/9781003619208-8

that often operates at the county, regional, or statewide level. Several local or "last-mile" networks, which serve homes and businesses directly, plug into the middle mile to gain access to the broader Internet. While last-mile networks can use a broad range of technologies (such as fiber-to-the-home, cable, and fixed wireless), middle-mile networks often use fiber optic. Remote rural communities often rely on a single fiber path to gain access to the middle mile. If the path gets damaged (for example, due to construction work), Internet service will be lost until it gets repaired. Colorado provides funding for projects that provide one or more backup cables to re-route data transmission and prevent interruptions – which is known as "network redundancy." The state is aware that last-mile grants must be supplemented with funding for redundant middle-mile infrastructure.

The state's approach is to run separate programs that target the middle and the last-mile networks. The Department of Local Affairs (DOLA) awards grants to public-private partnerships (PPPs) for middle-mile networks that provide service to the last mile and to anchor institutions, such as schools and public service buildings. DOLA's Broadband Program is funded through the Energy and Mineral Impact Assistance Fund, which provides assistance to communities that are economically impacted by the fuel and mineral industry. For last-mile funding, private internet service providers (ISPs) applied to the state's now defunct Broadband Deployment Fund, located within the Department of Regulatory Affairs (DORA). While the program prioritized projects in low-density areas, DORA did not require applicants to partner with a local government. The state allowed grantees to combine funding from DOLA and DORA. Today, last-mile funding is awarded by Colorado's Broadband Office.

Maine

Low population density and a challenging geography prevent rural communities in Maine from attracting private sector investment. The state is notable for its geographic isolation from the rest of the country; scant and dispersed population; sharp geographic and economic divides, and inclement weather. Maine communities have developed a strong tradition of local self-determination, interlocal cooperation, and regionalism (Palmer et al., 1992), which enables small communities to gain access to essential services. Broadband is no exception. The state acknowledges that the traditional private investment model is not always feasible (ConnectMaine, 2020), and thus supports both traditional and non-traditional models of investment, network ownership, and Internet service provision. PPPs between local governments and private ISPs ensure community support for private sector projects. Municipalities are allowed to partner with their neighbors and form broadband utility districts (BUDs), which are regional entities that can finance, own, and/or operate broadband networks. One of our case studies features Maine's first operational BUD, Downeast Broadband Utility (DBU).

Maine also requires that underserved communities applying for funds exhaust all traditional options first (ConnectMaine, 2020). Negotiating an expansion or an upgrade of the existing network with the incumbent provider (if there is one) is probably the quickest, easiest, and least expensive option. They already have the infrastructure, but it also means they have an advantage over competitors. As we show in this chapter, incumbents are more inclined to invest if there is local, state or federal funding available. Because ConnectMaine was a small program with a limited budget, early on the office partnered with providers to seek federal funding for rural communities. Today, there is more funding available due to a major contribution from the Legislature (Quaintance, 2021). ConnectMaine is now a unit of the Maine Connectivity Authority, a new office established in 2021 to manage ARPA and IIJA funds (Gray, 2023).

Middle-mile access is robust but not universal. For more than a decade, Maine has relied on individual, privately owned middle-mile networks scattered across the state, including the Three Ring Binder, which connects over a hundred communities and six hundred anchor institutions (Kittredge, 2013; Anderson, 2019). Today, the state is taking new steps to increase middle-mile access. In 2023, Maine received $30 million from the NTIA to build a middle-mile fiber network spanning 530 miles and connecting 131 communities (Maine Connectivity Authority, n.d.).

Policy barriers

When an agreement with an incumbent fails to materialize, rural communities are forced to consider alternative service models. State regulation and funding play a critical role in providing communities with options. However, to access state funding, communities must comply with a number of eligibility criteria and grant requirements – which can also become barriers. Table 6.1 shows the infrastructure, provider, and area eligibility criteria, and financial requirements of the three programs observed in our case studies. While rural communities in Colorado and Maine face similar geographic and fiscal constraints, they must navigate unique eligibility criteria, grant requirements, and broadband regulation to access state funds.

How could these state program policies become barriers? For one, provider eligibility requirements can harm underserved areas and non-traditional providers. While states understandably prioritize unserved communities, underserved communities are also in need of state support. Often, they are served by only one provider, and struggle with uneven access and slow, unreliable, and/or unaffordable Internet service. Without competition, there is little incentive for incumbent providers to expand or upgrade their services. Communities will then consider partnering with a non-traditional provider (such as an electric cooperative) or financing their own networks. While not all local governments pursue these alternatives, states can restrict their authority to do so. If restricted by state

TABLE 6.1 Selected Policies of Colorado and Maine Broadband Programs

State Policies		State Program		
		Colorado (DOLA)	Colorado (DORA)/ Colorado Broadband Office	ConnectMaine
Infrastructure eligibility	Infrastructure type	Middle mile	Last mile	Last mile
	Technology, download/upload speeds	Fiber optic is expected	No technology preference stated, but must deliver 25/3 Mbps or higher	No technology preference stated, but must deliver 10/10 or higher (increased later)
Provider eligibility	Applicant must be a public-private partnership	Required	Not required	Not required
	Electric cooperatives can apply	Yes	Yes	Not specified
	Municipal broadband projects are eligible	Yes (with referendum approval)[a]	Yes (with referendum approval)	Yes
Area eligibility	Area where an existing provider has already received federal funding is eligible	Not specified	Not specified.	State funds cannot be used in areas where a competitor already has plans to build
Financial requirements	Required match (%)	25%–50%	25%	25%

Sources: Colorado Department of Local Affairs, Colorado Broadband Office, ConnectMaine (2020), Cooper (202).

a *In areas subject to municipal broadband restrictions, DOLA funds can only be used for "dark fiber" if the project benefits non-governmental users. Middle mile "dark fiber" remains unused until leased to an Internet Service Provider.*

law, these projects are not eligible for state funding. One example is municipal broadband restrictions, which existed in Colorado until 2023. The state required public broadband projects to obtain referendum approval. By contrast, Maine imposes no restrictions on municipal broadband projects, allowing local governments greater autonomy in deciding how to address their connectivity needs.

Area eligibility requirements can also restrict access to state funding. When communities apply for state funding, Maine requires that they first try and negotiate with an existing provider. If negotiations fail, a community could still apply for state funding without partnering with the provider. However, an area is rendered ineligible if the provider has finalized plans to build there, and if the project meets the state's speed threshold (ConnectMaine, 2020). If the provider received federal funding, that signals their plans to build. However, state policy does not take into account that federal funding is awarded based on census blocks, not jurisdictions. Federal programs are guided by federal broadband availability maps, which also use census blocks. Communities can be segmented into several census blocks, and sometimes, only a few blocks are pre-awarded federal funding – those blocks that were included in the provider's application. With only a few pre-awarded blocks, a provider can discourage any competing projects. If the community wants to apply for state funding on their own, they will need to exclude these pre-awarded census blocks from their state grant application – which could make their project less financially viable to prospective partners, and to the state program.

Additionally, each federal and state program has its own requirements. What happens if the federal program's benchmark is outdated? Not so long ago, federal funds were still subsidizing new infrastructure that only delivered 10/1 Mbps (Neenan, 2023). This benchmark is slower than Colorado's (25/3 Mbps) and Maine's (10/10 Mbps). Should entire communities be precluded from applying to state funding if a few census blocks are already tied to a slower project?

Finally, burdensome grant requirements harm communities with limited resources. Communities might struggle to cover a percentage of the project's costs, which in turn limits their chances to secure a private sector partner. Both Colorado and Maine require grantees to match at least 25% of the project's costs.

This chapter shows how communities find a way around these barriers … or how these barriers can cause projects to fall apart.

From State to Local

Drawing from the Pew data base of state grants described in Chapter 3, we selected recipients that met the following criteria: (1) The locations were rural, or metro-adjacent; (2) the providers were local and, preferably, non-traditional rural telephone and electric cooperatives, and public network operators; (3) the technology delivered was fiber broadband, which is regarded as the gold standard of broadband speed, capacity and reliability. The cases illustrate how communities leverage public funds, navigate grant requirements, and pursue creative, place-based solutions, despite low population density and limited fiscal capacity.

We reached out to local government representatives, non-profit organizations, and broadband providers from communities in the two states. Thirteen interviews were conducted between October 2023 and January 2024 with key actors involved in these state-funded broadband initiatives. These include:

Emily Lashbrooke, Executive Director of Pagosa Springs Community Development Corp. in Archuleta County, CO;

Kent Blackwell, Chief Technology Officer of Delta-Montrose Electric Association (DMEA) in Colorado; Eric Jaquez, Operations Administrator of Rio Blanco County's Operations Department, CO;

Jim Fisher, Town Manager of Deer Isle in Hancock County, ME;

Butler Smythe, Organizer of the Peninsula Utility for Broadband (PUB) in Hancock County, ME;

Sarah Davis, Vice President for Market Development, Consolidated Communications in Hancock County, ME;

Josh Gerritsen, Former Chair of the Lincolnville Broadband Committee in Waldo County, ME;

Dan Sullivan, President of the Downeast Broadband Utility (DBU), and broadband advocate in Washington County, ME.

Insights from these interviews and other project documents, including local broadband planning documents and U.S. Census data, were used to develop the case studies which follow (Bravo and Warner, 2024; Bravo, 2025).

Maine: State Leadership, Policy, and Funding Expand Local Options

State policy in Maine allows both traditional and non-traditional approaches to broadband deployment. The state is aware that several rural communities are too small to pique the interest of private ISPs, or to finance their own broadband networks. The state's strategy is to provide communities with options. On the one hand, Maine allows multiple municipalities to join and form interlocal organizations called "Broadband Utility Districts (BUD)," which allow them to share the costs of building and operating broadband infrastructure. Historically, utility districts have been created to deliver essential services like water, electricity, and waste disposal to rural communities in Maine and Vermont. On the other hand, Maine also helps communities that prefer a traditional private provider, by awarding state grants to PPPs and assisting with applications for federal funding.

We profile both public and private fiber optic networks in rural communities. On the public side, we have Maine's first BUD, DBU in Washington County. The founding members, Calais and Baileyville, used a Line of Credit to finance the construction of an open-access fiber optic network. Four communities then received state grants to join DBU, and the revenue growth now helps with the repayment of the original loan. On the private side, we have the PUB, a coalition of seven towns from the Blue Hill Peninsula and Deer Isle, in Hancock County. The towns gained access to fiber after receiving attention from the state's broadband authority, which partnered with the region's incumbent provider to apply for federal funding. Finally, we have the Town of Lincolnville, in Waldo County, where the only provider has already deployed fiber – but the service is both costly and physically unavailable to many households. The town explored both traditional and non-traditional solutions – including negotiating with the incumbent, joining a BUD, and financing its own network. Due to financial and policy constraints, Lincolnville cannot move forward with any of these solutions.

In BUD's case, state funding allowed the original, loan-financed network to expand and provide fiber access in four new communities. In PUB's case, state leadership motivated the incumbent provider to upgrade their network to fiber. Lincolnville's case shows that state policy must be nuanced and flexible, to give communities leverage against monopoly power. When there is no competition, the incumbent does not have the incentive to expand or lower prices. When there is already fiber, the town is not eligible for state or federal funding – making it more difficult to build its own network, or join other towns and form a BUD. Below, we explore their stories in detail.

Downeast Broadband Utility: Going Down the Public Route

In 2015, Julie Jordan, Director of the Downeast Economic Development Corporation in the city of Calais, joined forces with Daniel Sullivan, then IT Manager at a local pulp manufacturer in Baileyville, to identify ideas that would bolster the region's economic development. Calais and Baileyville are located in Washington County, near the Canada-US border. Together, they are home to 4,600 people. After talking with town councils, local businesses, and schools, Jordan and Sullivan concluded that the region needed a robust fiber-to-the-home (FTTH) network. While the communities had access to cable, DSL, and satellite, residents and businesses were increasingly frustrated with the capacity and cost of the available Internet service.

Why fiber optic? Capacity is not the only thing that matters for Maine connectivity: resiliency to weather matters, too. The state is well-known for its heavy snowstorms and strong winds. Fiber networks "can withstand crazy weather," says Sullivan. Optical fiber strands are made of glass or plastic, and so the conduit places less pressure on poles than copper conduits do. Not long before our interview, a recent storm had pulled down telephone poles, and Internet service was out for hours. However, the fiber conduit, which was laying on the ground, was still working. Such is the faith of Daniel Sullivan in fiber optic, that years ago he lobbied for the state's broadband office to raise the minimum download/upload speeds from 3/1 Mbps to a symmetrical 10/10 Mbps. "By being in the room is when you can actually effect change. Especially if the only people in the room have been the incumbents," argues Sullivan.

Mindful of project costs and the region's low population density, Calais and Baileyville first considered a traditional private sector route. They approached Spectrum, and raised the question of building a fiber-to-home (FTTH) network that served every home in Calais and Baileyville. The communities were even willing to match a percentage of the project's costs – but received a flat-out refusal. Spectrum was wary that a public contribution would entail public network ownership, and preferred to retain full control of the network. Thankfully, Maine allows communities to explore other options.

The Role of State Policy

Calais and Baileyville then opted for the non-traditional public route – pooling their resources and financing the network construction themselves. Since 2015, the state has allowed the formation of BUDs, and two years later, Calais and Baileyville founded the first one in Maine, DBU.

> If a utility service is not being provided by private enterprise, then the local community has the authority to go ahead and do that. We didn't really have to leverage it too much. We just went in and said, "We're going to set up a nonprofit utility." We got no resistance from the state whatsoever, Daniel Sullivan says.

Neither Calais nor Baileyville was considered unserved, so they were unlikely to get state or federal funding. BUDs can issue bonds, but DBU opted to approach the local banks. As it turns out, the banks were directly affected by how slow and expensive Internet access was in the region, and were interested in financing the project. Instead of issuing bonds, DBU used lines of credit and customer fees to finance construction. In total, the project's costs amounted to $3.1 million.

Construction of the FTTH network began in 2018, and was completed in 2020. The network aimed to reach every home in Calais and Baileyville, and in some places, it has expanded as far as ten miles to reach a single home. DBU owns the network, and multiple providers can lease fiber from DBU and provide Internet service to homes and businesses. Currently, there is only one provider – Pioneer Broadband – but this works well enough for the district. Pioneer charges customers $67 every month for a 1 Gigabit connection (1,000 Mbps download, 1,000 Mbps upload), but can offer higher speeds as well. Out of that $67, $25 goes back to DBU and is used to pay for the network maintenance, the repayment of the loan, pole licensing, insurance, etc.

The Role of State Funding

Gaining more customers is key for faster repayment of DBU's loans – and that's when ConnectMaine, the state's then broadband authority, and the Maine Community Foundation, a non-profit, come in. Even though the original project was not eligible for state funding, public and non-profit dollars helped offset the costs of expanding the network. In 2021, the town of Alexander, thirteen miles away from Calais, received a $147,000 grant from the state and a $10,000 grant from the Foundation to join DBU. In 2021, the Passamaquoddy Indian Township Reservation received $315,000 to work with Pioneer Broadband, through the efforts of the Maine Community Foundation. ConnectMaine then awarded the Township a $105,056 grant to join DBU.

In 2022, the towns of Cooper and Princeton voted to join DBU, and later received $400,000 and $235,309 from ConnectMaine, respectively. Unlike Calais and Baileyville, Alexander, Cooper, and Princeton only had access to DSL, so they were not regarded as "served" and could apply for state funding. Except for Indian Township, the state grants do not cover the full extent of the expansion's costs, so DBU members will need to pool their resources together to finance the rest. But with state funding, expansion is now feasible, and the increase in revenue will allow DBU to repay the loans more quickly.

> The beauty of our model is that, once we have paid the loans, and we allocate enough money for maintenance and the upgrades we may possibly do, that's a revenue stream. That turns into a revenue stream for these communities […] The model that we use sends the public dollars back to the public, Daniel Sullivan concludes.

Ideally, because DBU members are collectively responsible for repaying all loans, new members should bring outside funding. This became a barrier for the town of Charlotte, thirteen miles south of Calais, when it sought to join DBU. A cable company had recently been awarded federal funding from the FCC's Rural Digital Opportunity Fund (RDOF) to build fiber in a portion of the town, which accounted for more than half (57%) of its population (Whelan, 2023). Since Maine does not fund areas where an existing provider already has plans to build, Charlotte would need to apply without that portion of the town. However, with half of its demand cut, the town was no longer a likely candidate for state funding, or a viable member for DBU.

Peninsula Utility for Broadband: Going Down the Private Route

Despite its name, the PUB is not a BUD. PUB is a coalition of five towns in the Blue Hill Peninsula (Blue Hill, Brooklin, Brooksville, Penobscot, and Sedgwick) and two towns on the island of Deer Isle (Stonington and Deer Isle). The towns are located in Hancock County, a hill-and-valley region whose major industries include commercial fishing, farming, and tourism. The Blue Hill Peninsula is surrounded by a rugged coastline and blueberry barrens, and the largest town, Blue Hill, has a population of only 2,800 people. There is a large blue-collar workforce, but many of the region's homes are also holiday properties maintained by wealthy out-of-towners.

The region's access to fiber optic broadband is only a recent phenomenon. The towns originally had uneven access to DSL, satellite, and cable – with large pockets lacking access of any type. While cable offered adequate access, only a few towns had access to it. Customers had to pay thousands of dollars for line extensions between the cable network and their homes. Few were willing to make such an investment – not only are many homes only for seasonal use, but many year-round households lacked the means, as well.

In 2018, the towns of the Blue Hill Peninsula and Deer Isle began to explore various options to bring affordable fiber to their constituents. They formed PUB, which was originally intended to be a BUD, but was never formally organized as such. Four PUB towns began exploring the option of issuing bonds to finance a regional FTTH network, and sent a Request for Proposals to several ISPs. While they received nine bids, no plans ever materialized because another option unexpectedly presented itself: One of the providers that bid on the Request for Proposals, Consolidated Communications, partnered with ConnectMaine to go after federal funding, and included all seven PUB towns in their application.

Jim Fisher, the Town Manager of Deer Isle, regards the coalition as critical to pique the interest of private providers like Consolidated. While the towns themselves did not apply for funding, they pooled together their demand and gave the impression of a larger market.

> My general philosophy is that our small towns don't do very well on our own … For any investor, it's much more attractive to go into an area with a population of 10,000, than a population of 700. For them to put down all of that infrastructure… They have to have an area big enough for it to make sense. … They don't want to go in and find they've got to do this sort of leap-frog installation. By organizing and working together, I think we were big enough to get some attention. We appeared bigger than we were.

The Role of State Leadership

When the NTIA announced the round of applications for its National Broadband Infrastructure Program, ConnectMaine's strategy was to partner with three of the largest ISPs in the state, including Consolidated Communications. These providers had the resources to match a substantial percentage of the project's costs, which in turn would make the application more competitive in the eyes of the NTIA. The strategy was successful, and in 2021, the state received $28.09 million for seven projects that would connect 11,746 unserved households. The Blue Hill Peninsula/Deer Isle project received $9.83 million for the construction of FTTPs in eight locations.

> We [Consolidated] had been working with the state of Maine for years on broadband expansion. The state's biggest challenge … was that it never had

any significant funding sources for such things. ... The ConnectMaine Authority was [then] led by Peggy Schaffer. She immediately recognized that we wanted to go after this [NTIA] funding, and she wanted to win. And part of that was putting together a project that fit really well into the [federal] requirements, says Sarah Davis, VP for Market Development at Consolidated Communications.

In many ways, Consolidated was the logical choice for PUB. Consolidated already owned infrastructure in the region – copper cables that hung on poles. In many places, they mainly needed to attach the fiber cables to the poles. Any other provider would need to start from scratch. Second, Consolidated and its subsidiary, Fidium Fiber, had already received funding from the FCC's Rural Digital Opportunity Fund to build fiber in portions of the Blue Hill Peninsula. While they had not started building yet, any other provider seeking state or federal funding would need to apply without those portions of the town – which, in turn, would make the project less profitable. Finally, Consolidated had already built fiber in part of the town of Stonington, in Deer Isle. Years before, Stonington had negotiated with Consolidated to upgrade its DSL network. However, instead of faster DSL, Consolidated decided to deploy fiber. Stonington contributed $200,000, and was the first town in the region to gain fiber. From there, it only made sense for Consolidated/Fidium to expand. Fidium Fiber has laid fiber across the Blue Hill Peninsula and Deer Isle. Installation is free for subscribers, with monthly fees starting at $25/month for 100/100 Mbps.

There is more than one approach to gaining fiber, and PUB towns disagreed on the approach. Two of PUB's original members were adamant that the network should be publicly owned and operated, and temporarily broke away from the coalition. This was a blow for the coalition, as the project was more attractive to Consolidated if there were larger demand. Public ownership was not truly a priority for the other PUB towns, who were mindful of the time and resources required to run a public utility. In addition, their constituents were against using tax dollars to build a network. They would have had to apply for federal funds ... which Consolidated had already done, and successfully so. The project was fully funded by the NTIA's grant and capital from Consolidated, so the towns did not have to raise taxes or issue bonds to provide a local match. Finally, if PUB towns wanted to build their own network, they also would need to lay new conduit between the mainland and Deer Isle. Consolidated already had a cable running underwater with enough capacity to support user traffic in Stonington and Deer Isle. The coalition did not see a way in which they could have competed.

PUB did not establish a BUD in the end, but they were able to gain fiber thanks to interlocal cooperation and state leadership. Jim Fisher says, half-jokingly: "Maybe in the long run, we'll regret in some way. They'll use their monopoly power and squeeze money out of us. But for now, at least … It's working out really well."

Lincolnville, Waldo County: When Incumbents "Capture" Underserved Communities

The town of Lincolnville, with approximately 2,300 residents, is unique among the case studies in this chapter for one reason: There is already fiber in the town. Lincolnville is solely served by one company, the Lincolnville Telephone Company (LCI), through its subsidiaries, LCI and Tidewater Telecom. The town's residents were frustrated with LCI's monopoly of the town's communications services, as subscribing to Internet service was expensive. Customers were required to enter into three-year-long contracts with costly subscription fees and unnecessary services that they could not opt out of, like phone lines. As for LCI's fiber optic Internet, the company had only deployed fiber in some areas of the town, and offered relatively low download/upload speeds for a high price. To illustrate: In 2022, LCI charged $64.95 for 50/10 Mbps, while DBU in Washington County charged $67 for 1 Gbps (1,000/1,000 Mbps).

In 2019, LCI approached Lincolnville with plans to expand their fiber optic network across the entirety of the town. They offered to expedite the process if the town provided a local match, which would give LCI a better chance of securing a state grant. However, the company only meant to deploy fiber along all the major roads – not connect every home. There was also no chance of co-ownership, which the Lincolnville Broadband Committee considered an imperative if public funds were used. The Committee declined to support the project, and sent a letter to inform ConnectMaine of their refusal.

Barriers to Going Public

Since negotiations with LCI had gone nowhere, Lincolnville then explored a number of options to improve their fiber connectivity. In 2021, the town decided to join the Midcoast Internet Coalition, a group of municipalities in the Midcoast region that planned to establish a BUD and finance the construction of an open-access fiber optic network. Some of these towns already shared the costs of other essential services, such as transportation and an emergency call center. By banding together, they hoped to increase their chances of receiving state funding as well. However, while some of the coalition's members moved forward with these plans, Lincolnville chose to opt out. To ease the costs of deployment, federal or state funding is critical – but these programs prioritize unserved locations, and LCI's fiber in parts of Lincolnville rendered the entire town ineligible for funding. Any competitor going after state funding would have to exclude parts of the town where LCI had fiber – and yet, without these parts, the project would not be profitable for anyone else but LCI.

As with PUB and Consolidated/Fidium, LCI was essentially Lincolnville's only viable choice. Why did things go south in Lincolnville? For Josh Gerritsen, the former Chair of the Lincolnville Broadband Committee, it was the issue of giving away public dollars without getting public ownership.

> The only real partnership that was realistic for us was working with LCI and putting in a matching payment. We would have no ownership of it. Because you don't, the private company will also have all the pricing powers in the future. That was one of the bargaining chips the town hoped to have. If we put in a matching fund, we would insist on a pricing cap.

Lincolnville also did a feasibility study to evaluate the potential for a dedicated network for the town. Alas, the study concluded that Lincolnville was unlikely to receive any state or federal funding, and that it would be difficult to defend going into debt to deploy new fiber in a town that already had fiber access in some areas. In the end, Lincolnville's public broadband efforts were stalled. While prices remain high, LCI has begun to offer higher speeds, and has dropped the three-year contracts and phone line requirements.

Gerritsen says that local connectivity efforts must be timely, or else they will be left with no options. Incumbents can expand their networks without trying to reach every home. If they meet the state's definition of "served," then the area is unlikely to get state funding. Since fiber is available in Lincolnville, the incumbent can challenge any grant application from a competitor. This has essentially facilitated Lincolnville's monopoly capture:

> If your town doesn't have much fiber deployment, that is the opportunity to potentially build out your own system. I mean, it's not too late at that point, you know, in terms of getting a low interest loan from a bank or from the state. Because once you have that incumbent with fiber up there, they're going to say, "Well, there's an incumbent. How can you justify this huge loan? How are you going to get the take rate that's going to pay it back?"

When it comes to broadband access, the state is meant to provide underserved rural communities with options. The state's BUD legislation allowed Baileyville and Calais to form the DBU, and state funding facilitated the network's expansion. State leadership was instrumental in getting Consolidated to deploy fiber in the Blue Hill Peninsula and Deer Isle, and the towns did not have to raise local funds, which suited them better. At the same time, Lincolnville's case shows that state funding is critical if a community wants to explore non-traditional options like building their own network, or sharing the cost with their neighbors. Public ownership is especially important for communities like Lincolnville, who are essentially "captured" by the incumbent provider and struggle with their current service. Even if they do not opt for a public network, having several options provides underserved communities with leverage when negotiating with providers. States will need to consider how they can use policy and funding to empower rural communities, and center public interests over private profit in broadband deployment.

Colorado: State Support for the Middle and Last Miles Creates a Robust Foundation

Referring back to our multi-level governance framework (Figure 1.1) in Chapter 1, state broadband programs play a crucial role in the distribution of funding to providers and localities, which helps address rural challenges. Colorado is well aware that improving rural connectivity is not only a last-mile problem – but a middle-mile problem as well. For almost a decade, Colorado's DOLA has subsidized the construction of redundant and high-capacity middle-mile networks. Rural communities in Colorado navigate state policy and leverage state funding to build resilient broadband networks and bolster last-mile competition. Colorado recognizes that improving rural connectivity takes a regional approach. State policy must enable the development of strategic partnerships between local governments and non-traditional providers. While not lifted until 2023, Colorado's municipal broadband restrictions were relatively light compared to other states. Municipalities were able to opt out of these restrictions through a referendum vote, and we found examples of broadband networks that are owned and operated by local governments, non-profits, and electric cooperatives.

In Archuleta County, the aim was to prevent the frequent Internet outages that plagued Archuleta and its neighbors – so the county opted to invest in a redundant middle-mile network. First, state funding was used to build the county's first redundant fiber loop. "Network redundancy" means that, if one of the network's fiber cables is damaged, user traffic can be maintained by switching to the second cable. Second, Archuleta partnered with La Plata County and the Southern Ute Indian Tribe and applied for state funding to build a second fiber path. This path will serve as an alternative to the existing fiber cable coming from La Plata County, where service often gets cut due to environmental factors or construction accidents.

In Rio Blanco County, the aim was to bolster last-mile competition – so the county decided to take care of the infrastructure themselves. Rio Blanco's open-access last-mile fiber network is owned and operated by the county, funded with county and state funds, and developed in partnership with a local electric cooperative. Finally, in the counties of Delta and Montrose, residents struggled with the costly and unreliable service offered by large private companies. The DMEA, a member-owned cooperative, saw an opportunity to compete. They had already built an internal fiber network to connect all of their substations, which serves as their high-capacity middle-mile network. DMEA leveraged state funding to provide last-mile service in Delta and Montrose.

Archuleta County: Using a Regional Mindset to Prevent Internet Outages

Archuleta County, located in southwestern Colorado, spans over 1,355 sq. miles and has a population of 14,189. The town of Pagosa Springs, its only incorporated municipality, is relatively isolated, as it is largely surrounded by the San Juan National Forest. The national forest covers over half of the county's territory and spans into the Southern Ute Indian Reservation. The town and the county co-own some underground fiber in the area, and outsource the management of their broadband assets to a non-profit called Pagosa Springs Community Development Corporation.

Today, Pagosa Springs CDC is responsible for developing the county's middle-mile fiber network and working with various ISPs. Their closest partner is Visionary Broadband, an ISP that provides 1 Gigabit Internet service in downtown Pagosa Springs. In 2019, Visionary secured a $466,000 grant from Colorado's Department of Regulatory Affairs for a last-mile project in Archuleta and its neighbor, Hinsdale. The project involved deploying cell towers to provide fixed wireless service for 577 households in remote areas where fiber is cost-prohibitive.

Leveraging State Funding

Archuleta residents struggle with frequent Internet outages. The county is served by a single fiber path coming in from La Plata County. When service gets cut, due to environmental factors or construction accidents, this has a domino effect on La Plata's neighbors, the Southern Ute Indian Tribe and Archuleta, who lose Internet service as well.

> We lose cellphone signal, we lose landlines, we lose emergency management… We lose everything because a neighboring community cuts a line. … It has happened three times this year already. And it cripples us, because we are up against a mountain. We're rural Colorado. We're a large county, describes Emily Lashbrooke, Executive Director of Pagosa Springs CDC.

The county's strategy was to build resiliency at the local and regional levels. First, to address internal fiber cuts, Pagosa Springs CDC used DOLA funds to build a redundant fiber loop that connects Carrier Neutral Locations, which allow ISPs to interconnect with the local middle-mile network. In a loop, if one fiber cable gets damaged, data transmission can be preserved by switching to the other cable.

Second, to address regional fiber cuts, Archuleta, La Plata and the Southern Ute Indian Tribe partnered with the La Plata Electric Association, a local electric utility, and Region 10 League for Economic Assistance, a non-profit that provides middle-mile fiber access in western Colorado. The idea was to deploy a second fiber path running between the town of Ignacio in La Plata and Pagosa Springs in Archuleta. Ignacio is part of the reservation, where there are miles of fiber optic infrastructure that can be leveraged for a regional network. The Tribe consented to grant ownership of part of its fiber to Archuleta, La Plata, La Plata Electric, and Region 10 for 20 years. In exchange, the Tribe is to receive $4 million – half to be covered with a $2 million grant from DOLA, and the other half with four contributions of $500,000 from each of the parties (Schafir, 2024).

State funding has been critical to the development of resilient fiber optic infrastructure in Archuleta. The county's success at securing state funding can be partially attributed to their strategic partnerships with their neighbors, electric cooperatives, and non-profits. By pooling their resources together, they can craft solutions that have not only a local, but also a regional impact.

> A regional mindset is a good [thing] to look at. Talk to your neighbor, find out what they are doing. You don't have to reinvent. Leverage any funding you have. Go after [funding], because it's out there, right now. There's more money than we're ever going to see again in our lifetime. Leverage. Partner, suggests Emily Lashbrooke.

Delta and Montrose: The Role of Electric Utilities in Broadband

Delta-Montrose Electric Association (DMEA) is a local electric cooperative that operates in five counties in western Colorado, but primarily serves the counties of Delta and Montrose. There are several large telephone and cable providers operating in the region, yet Internet service remained expensive and unreliable for residents. Traditional providers were not interested in upgrading their infrastructure, so residents turned to a non-traditional provider instead. In 2015, after becoming aware that DMEA was building an internal fiber optic network to connect its electric substations, the cooperative's member-owners saw an opportunity to compete with ISPs.

> Our members rose up in awareness of that project, … pleading with the co-op to get into the broadband service space. They were just tired of high-cost providers, poor reliability, just a disinterest to really improve their services in any meaningful way. The members stormed one of our board meetings … and just pleaded their case, relates Kent Blackwell, Chief Technology Officer of DMEA.

The following year, DMEA established its own broadband subsidiary, Elevate Internet. As an electric cooperative, it has some unique advantages. For one,

DMEA/Elevate already has access to electric easements and utility poles, which speeds up the process of installing aerial fiber. The cooperative can focus exclusively on the last mile, as the fiber network they already built works as their high-capacity middle mile infrastructure. In fact, DMEA leases excess fiber capacity to other middle-mile providers.

Since its foundation, DMEA/Elevate has been proactive in going after both state and federal funding. The cooperative has received funding from the Department of Regulatory Affairs for a number of last-mile projects in western Colorado, including two grants to build in Delta County ($759,585 and $2,334,988, both in 2018), and two grants to build in Montrose County ($683,158.58 in 2019 and $1,431,083 in 2020).

Rio Blanco County: Leveraging Public Ownership to Bolster Competition

Rio Blanco is a sparsely populated county in northwestern Colorado, with only 6,500 residents scattered across 3,221 square miles. The largest settlements are the towns of Meeker and Rangely, and most of the county's territory is public land controlled by the Bureau of Land Management. Rio Blanco's vastness and remoteness provided a thought challenge for ISPs, who still utilize old infrastructure (such as telephone copper lines and radio waves) to transmit data. Older technologies can no longer meet the current demand for high-speed Internet service, but there is no incentive for providers to upgrade their infrastructure. This upgrade also would involve connecting to the nearest middle-mile network, which makes projects in remote counties like Rio Blanco even more expensive. The county realized that if they wanted fiber connectivity, they would need to provide the infrastructure themselves.

For more than a decade, the town of Meeker's anchor institutions (school district, county hospital, and local library) had been connected by fiber. Rio Blanco planned to deploy FTTH infrastructure in Meeker and Rangely, and to deploy a tower network to support fixed wireless service in the more rural parts of the county (Rio Blanco County, 2014). The project was pushed by Rio Blanco County's IT Director, Blake Mobley, who previously worked as the school district's IT Director. Mobley was inspired by Google Fiber's $70-per-month 1 Gigabit service model. For Mobley, the price was relatively affordable. Despite the substantial deployment costs, fiber was also a sensible long-term investment, as maintenance and upgrade of a fiber network are relatively inexpensive.

Navigating State Policy, Leveraging State Funding

The county had not considered going after state or federal funding, but then an opportunity presented itself. In 2014, Rio Blanco overrode Colorado's municipal broadband restriction, with an 82% referendum approval. That same year, Colorado's DOLA established its broadband initiative, to support middle-mile deployment and regional broadband planning. Once DOLA announced the first round of applications for its Broadband Program, the county was able to move fast.

In 2015, Rio Blanco was awarded $2 million by DOLA. The program does not fund last-mile projects but, grant recipients can use DOLA funds to cover the costs of building the middle-mile component of the project. Rio Blanco would need to cover the rest of the project's costs with local funds, but it was better positioned to do so than most. The county had several revenue sources from oil, natural gas, grazing, ranching, and the use of public land. The county also benefited from owning several of the region's roads and rights-of-way, which sped up the permitting process. To gain access to electric easements and utility poles, the county partnered with a local electric cooperative, White River Electric.

Access to a high-capacity middle-mile network was also critical for Rio Blanco. The county's network connects to Project THOR, a middle-mile network that serves fourteen rural communities in Northwestern Colorado. Project THOR built a series of redundant fiber loops that provide communities like Rio Blanco with two fiber paths. The initiative leveraged fiber that was already installed, and funds were mostly spent on connecting individual fiber networks, including Rio Blanco's. Project THOR was coordinated by the Northwest Colorado Council of Governments, and received funding from DOLA and local matches from the communities that participated in the project (Chuang, 2020).

From Network Owner to Internet Service Provider

Over the years, Rio Blanco has seen some unexpected changes to their initial plans. The county originally intended to build the infrastructure, and outsource its management, but remains the network's operator to this day. The county also expected multiple providers to use the network and provide Internet service, thereby providing county residents with options. As it turns out, low profits led to the exit of one of the two ISPs which leased fiber from the county. Rio Blanco then stepped in to provide residents with a second choice, and began providing Internet service directly. The majority of customers are subscribed to the other ISP, but Rio Blanco remains committed to maintain an open-access network and to promote competition. The network is operating with tight profit margins, and no longer has the resources to expand to the more remote areas of the county. Rio Blanco hopes that funding eligibility rules will become more flexible in the future, so state aid can once again help them serve underserved communities. Rio Blanco still has some way to go before achieving universal service.

Regionalism and Flexible Rules Are Key

In both Colorado and Maine, rural connectivity faces common barriers – low population density, remoteness, challenging terrains, access to slow and obsolete technologies, disinterest from incumbents to upgrade their services, costly Internet service, and unserved homes in "served" communities. Many of the communities featured in this chapter had Internet access, but were not truly served. This chapter illustrates how local actors took steps to bring fiber into their communities, the strategic partnerships that made these projects possible, how public broadband funding was leveraged, and how they navigated state and federal policies.

Three major themes stand out. First, a regional mindset proved essential in easing the path toward state funding, and tackling challenges of regional connectivity. Communities in Colorado and Maine demonstrate how interlocal cooperation and regional coalitions can significantly improve access to state funding, attract providers, and ultimately enhance broadband infrastructure in underserved areas. By pooling resources and aligning their efforts, rural communities can overcome geographic and financial limitations and increase their chances to access state funding.

Second, state eligibility criteria and requirements can become barriers to rural communities. These criteria can be weaponized by providers to prevent underserved communities from exploring other alternatives. Several of the communities profiled in this chapter were underserved, and did not qualify for state or federal funding. Without funding, communities have no leverage to negotiate with providers, or attract competitors. Providers, which have no intent to expand or upgrade, can still block applications from competitors. More flexible and nuanced funding rules are needed to prevent the "capture" of rural communities by unscrupulous providers.

Finally, state policy can empower rural communities to explore different network financing and ownership options. State broadband programs can support different approaches to rural connectivity, both traditional private and non-traditional public and cooperative. With state assistance, rural communities have been able to prioritize universal access, service quality, and affordability.

Bibliography

Anderson, J. C. (2019, May 17). *Despite conflict, new overseer of Maine broadband network promises to play fair.* Portland Press Herald. https://www.pressherald.com/2019/05/17/despite-conflict-new-3-ring-binder-overseer-promises-to-play-fair/

Archuleta County Broadband. (n.d.). About – Archuleta County Broadband. Retrieved January 17, 2024, from https://archuletacountybroadband.com/

Archuleta County Broadband Services Management Office. (2022). Strategic Plan 2022–2024. https://archuletacountybroadband.com/wp-content/uploads/2023/03/ArchuletaCounty_BSMO_StrategicPlan_2022-25_Final.pdf

Ban, C. (2023, November 6). Middle mile can be a matter of life and depth. National Association of Counties. https://www.naco.org/news/middle-mile-can-be-matter-life-and-depth

Bravo, N. (2025). *Multi-level Governance for Broadband Planning: Implications for State Policy and Interlocal Cooperation.* Unpublished PhD dissertation, Cornell University, Ithaca, NY.

Bravo, N., & Warner, M. E. (2024). Innovative state strategies for rural broadband: Case studies from Colorado, Minnesota, and Maine. Department of City and Regional Planning, Cornell University. Retrieved from https://labs.aap.cornell.edu/node/882

Census Reporter. (n.d.). Census profile: Lincolnville town, Waldo County, ME. *Census Reporter.* Retrieved January 17, 2024, from https://censusreporter.org/profiles/06000US2302739755-lincolnville-town-waldo-county-me/

Chuang, T. (2020, April 16). Internet service in western Colorado was so terrible that towns and counties built their own telecom. *The Colorado Sun.* https://coloradosun.com/2020/04/16/internet-service-western-colorado-rural-broadband-nwccog-sb152/

Colorado Broadband Office. (n.d.). Broadband Deployment Board & Fund. Retrieved April 26, 2024, from https://broadband.colorado.gov/broadband-deployment-board-fund

Colorado General Assembly (2005). "Concerning Local Government Competition in the Provision of Specified Communications Services." *Session Laws 2001-Present.* 2165. https://scholar.law.colorado.edu/session-laws-2001-2050/2165

ConnectMaine. (2020). State of Maine Broadband Action Plan. https://www.maine.gov/connectme/sites/maine.gov.connectme/files/inline-files/Plan_Action_2020.pdf

Cooper, T. (2024). Municipal Broadband Remains Roadblocked In 16 States. *BroadbandNow.* https://broadbandnow.com/report/municipal-broadband-roadblocks

Gray, M. (2023, January 26). Maine to see $34M push to connect rural areas. *GovTech.* https://www.govtech.com/network/maine-to-see-34m-push-to-connect-rural-areas

Kittredge, F. (2013) Maine's Three Ring Binder, Maine Policy Review, 22(1), 30-40. https://digitalcommons.library.umaine.edu/mpr/vol22/iss1/7

Local Government Provision of Communications Services, SB23–183, Colorado General Assembly 2023 Regular Session (2023). https://leg.colorado.gov/bills/sb23-183

Maine Community Foundation. (n.d.). Forward: Report to the Community 2020–2021. Retrieved January 17, 2024, from https://www.mainecf.org/wp-content/uploads/2021/07/2020-2021-AR-for-web.pdf

Maine Connectivity Authority. (n.d.). Connect the Ready Grant Program. *Maine Connectivity.* Retrieved January 17, 2024, from https://www.maineconnectivity.org/connect-the-ready-grants

Maine Legislature (2022). An Act to Support Municipal Broadband Infrastructure through Incentives and Competition. https://www.mainelegislature.org/legis/bills/getPDF.asp?paper=SP0664&item=3&snum=130

Neenan, J. (2023, November 9) 'It Was Graft': How the FCC's CAF II Program Became a Money Sink. Broadband Breakfast. https://broadbandbreakfast.com/it-was-graft-how-the-fccs-caf-ii-program-became-a-money-sink/

Palmer, K.T., Taylor, G.T., Librizzi, M.A. and Lavigne, J.E. (1992) *Maine Politics and Government*. Lincoln, NE: University of Nebraska Press.

Quaintance, Z. (2021, February 25). ConnectMaine expands affordable broadband across the state. *GovTech*. https://www.govtech.com/workforce/ConnectMaine-Exapands-Affordable-Broadband-Across-the-State.html

Rio Blanco County (2014) "Rio Blanco County's Request for Proposals for: The Rio Blanco County Broadband, Cellular, and ES Infrastructure and Services.". https://cdola.colorado.gov/sites/dola/files/documents/Rio%20Blanco%20County%20Broadband%20Services%20RFP.pdf

Schafir, R.M. (2024, January 25). Counties, Southern Utes leverage over $70 million in grant funds for broadband – The Tri-City Record. *Tri-City Record*. https://www.tricityrecordnm.com/articles/counties-southern-utes-leverage-over-70-million-in-grant-funds-for-broadband/

The Archuleta County Planning Commission. (2017). Archuleta county community plan. https://www.archuletacounty.org/DocumentCenter/View/2051/Archuleta-County-Community-Plan-2017?bidId

The Pew Charitable Trusts. (2021, November 29). How "Open Access Middle-Mile Networks" can facilitate broadband expansion. *The Pew Charitable Trusts*. https://www.pewtrusts.org/en/research-and-analysis/speeches-and-testimony/2021/11/29/how-open-access-middle-mile-networks-can-facilitate-broadband-expansion

The Pew Charitable Trusts. (2023, January 5). Vermont takes a regional approach to rural broadband expansion. https://pew.org/3Vdy0IM

WGME. (2021, May 10). Midcoast town coalition proposes local control over broadband services. *WGME*. https://wgme.com/news/local/midcoast-town-coalition-proposes-local-control-over-broadband-services

Whelan, L. (2023, February 10). Communities pursue options in face of broadband setbacks. *The Quoddy Tides*. https://quoddytides.com/communities-pursue-options-in-face-of-broadband-setbacks.html

7

GETTING AROUND PREEMPTION

The Power of Public-Private Partnerships

Duxixi (Ada) Shen, Mildred E. Warner, and Natassia A. Bravo

Preemption Undermines Local Initiative

In 2008, the city of Wilson, North Carolina, launched its own fiber optic network, Greenlight Community Broadband. Two years later, the small town of Pinetops expressed interest in getting Internet service from Greenlight. This was not an unusual request, as the city's public electric utility already provides energy service to Pinetops (Handgraaf, 2017). These plans were immediately stopped by North Carolina's new municipal broadband restrictions, passed in 2011.

The "Local Government Competition Act" states that any municipality interested in providing communications service to an underserved area will need to submit a petition to the North Carolina Utilities Commission, and give proof that the area is indeed underserved. Any private service provider or interested party will be able to object to the petition if the area is not underserved, or if the municipality is not qualified to be a provider. More critically, the law also prevents municipalities from providing communications service outside of their own jurisdictional boundaries (North Carolina HB 129, Session 2011).

In 2014, Wilson requested the Federal Communications Commission (FCC) preempt HB 129, arguing that Greenlight should be exempt (Murawski, 2014). Wilson was not alone – the FCC also received a petition from the city of Chattanooga, TN, which was engaged in a similar power struggle with their state (Holmes and Al Idrus, 2014).

Back in 1996, Chattanooga's electric utility company, the Electric Power Board (EPB), had built a fiber network to make their grid more resilient, and hoped the existing fiber would encourage communications providers to build the last mile and connect homes to the fiber network. When these hopes did not materialize, EPB gradually began providing retail Internet service to the city

DOI: 10.4324/9781003619208-9

at competitive prices. By 2009, they were serving 17,000 households. In 2010, EPB introduced 100/100 Mbps speeds, and then one year later, EPB announced their 1 Giga service (1,000/1,000 Mbps). Comcast filed a lawsuit to stop the EPB's plans (Mitchell, 2012). Back then, 1 Gbp was even rarer than it is today. After all, it was only in 2024 that the definition of broadband was updated to 100/20 Mbps, and 25/3 Mbps was the standard for almost a decade.

Like Greenlight, EPB intended to expand their service to neighboring areas. However, a 1999 state law prevented municipalities from providing communications services beyond their electric service areas (Holmes and Al Idrus, 2014).

In 2015, the FCC voted to have North Carolina's and Tennessee's laws overturned. Wilson was then able to expand into Pinetops (Gonzalez, 2018) – but this did not last. One year later, North Carolina and Tennessee successfully challenged the FCC's decision. The courts ruled that the FCC did not have the authority to overrule state law. Greenlight was only allowed to serve Pinetops until a private alternative emerged in 2018. Greenlight was then told to cease their operations and sell their infrastructure in Pinetops (Gonzalez, 2018).

Pushback Against Preemption

Over the years, there have been attempts to lift restrictions against municipal broadband in North Carolina and Tennessee. At least 22 states preempted municipal broadband before the pandemic (Cooper, 2024). Recent Federal legislation (e.g., ARPA, BEAD) specifically allows municipal broadband, but as of 2024, at least 16 states continue to maintain these roadblocks (Cooper, 2024).

Though it is hard to define clearly, state preemption is not new. Over the years, the manner in which states have limited local government decision-making power has grown and many municipalities complain of limits to their ability to serve the needs of their residents (Bravo et al., 2020). From simple restrictions, preemption has grown to be punitive against local leadership, and often reflects corporate power over state legislatures (Goodman et al., 2021). The National League of Cities (NLC, 2018) defines preemption as "... the use of state law to nullify a municipal ordinance or authority," and we see the impact of preemption across the cases we cover in this chapter.

Where private broadband providers are unwilling or unable to deliver the service, municipalities are investing in their own fiber optic networks, or request access to neighboring municipal broadband services. However, municipal leadership in the market has spurred the rise of state preemption of local authority to deliver and regulate municipal broadband. Municipal broadband is regarded as unfair competition by the industry, and a waste of taxpayer dollars by conservative state legislators. State legislation can curtail traditional local authority to own and operate broadband networks.

Preemption laws make it difficult for cities to acquire a service, improve their existing service, or to share the cost of services with their neighbors.

Additionally, local governments' ability to finance the construction of their own network, should they have the authority to do so, can still be impaired by state fiscal restrictions. Therefore, State preemption of local fiscal autonomy (through tax and expenditure limitations) can impact local investment in infrastructure and limit the expansion of access to services in low-income neighborhoods and rural areas (Bravo et al., 2020; Wen et al., 2020).

Historically, cities have been able to exercise control over the entry of capital and give away franchise and monopoly power. The level of city power has long been contested by business coalitions (Frug, 2001). The private sector welcomes state intervention, which is perceived as less restrictive than municipal oversight. The state level is also potentially easier to influence (Kim and Warner, 2018). Unsurprisingly, private telecommunications companies have spent millions fighting against municipal broadband (Common Cause, 2021; Block, 2022).

As of 2024, municipal broadband is explicitly restricted in 16 states, and faces barriers in another three. Some of these laws include: (1) Prohibiting cross-subsidization to fund broadband service; (2) Only allowing locations with municipal electric utilities to offer broadband; (3) Preventing service delivery outside of the public entity boundaries; (4) Requiring referendum approval; and (5) Restricting service to "unserved areas," with the municipality bearing the burden of proving that the area is indeed unserved and would not be served by a private provider (Chamberlain, 2023; Cooper, 2024).

Figure 7.1 shows the 22 states with roadblocks or explicit restrictions on municipal broadband before 2024. According to Cooper (2024), only 16 states

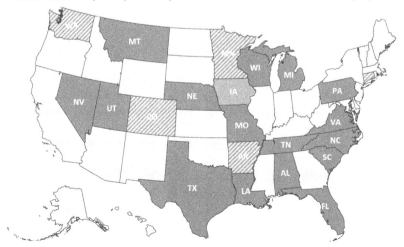

FIGURE 7.1.   Which states restricted municipal broadband before 2024?

Source: Author Analysis based on Chamberlain (2023) and Cooper (2024).

maintained explicit restrictions in 2024. He also lists three additional states that do not preempt municipal broadband, but have potential barriers – Iowa, Oregon, and Wyoming. We include Iowa in our map, as it is the only one of the three states that requires referendum approvals to establish public utilities and to issue municipal bonds to finance networks. The map also features five states that recently lifted their restrictions: Arkansas and Connecticut in 2019, Washington in 2021, Colorado in 2023, and Minnesota in 2024 (Chamberlain, 2023; Cooper, 2024). In Chapters 3, 5, and 6, we discussed how these laws impact state funding and community-led broadband projects in Minnesota and Colorado.

Local Initiative Despite Preemption

Local governments don't stop in the face of preemption. They look for other alternatives. We learned in Chapter 4 about the efforts of Brownsville, Texas, which partnered with Lit Communities, to circumvent the state's preemption of municipal broadband. Texas state laws restrict municipalities from offering specific telecommunication services directly by Utilities Code § 54.201. However, after another Texas city, Mont Belvieu, obtained local district court permission to build and operate its municipal broadband network, Brownsville saw an opportunity. It applied parts of the Mont Belvieu ruling when planning its broadband project. Brownsville funded a feasibility study, which revealed that many residents were paying high prices for substandard service and lacked affordable, high-speed broadband. This assessment, coupled with the effects of COVID-19 led the city commission to allocate significant ARPA funds to jumpstart broadband expansion. The ARPA funding was a transformative opportunity, providing initial capital to leverage further funding and quickly start construction.

After a competitive bidding process, to which the incumbent ISP did not even apply, the city partnered with Lit Communities, a mission-driven company that leverages public-private partnerships (PPPs) to deploy last-mile fiber optic broadband infrastructure. Brownsville is contributing $19.5 million of its ARPA funds, while Lit Communities is committing an additional $70 million. The city's portion will fund a publicly owned, open-access middle-mile fiber network, generating revenue via leases to potential ISPs. Lit's portion will establish private fiber-to-the-home (FTTH) connections for all residents and businesses. Upon completion in three years, Brownsville will have 100 miles of public open-access fiber backbone and 550 miles of private FTTH connections. The partnership with Lit Communities offers multiple benefits, including accessing more funding, reducing operational costs while maintaining control, and generating revenue. In addition, Lit Communities has committed to hiring local workers and providing additional services to address affordability and adoption.

ARPA opened up a window of opportunity for municipal broadband. Local authorities facing preemption were able to leverage ARPA funding to attract private partners and build publicly owned broadband networks. The recent Broadband, Equity, Access, and Deployment (BEAD) Program could also challenge the restrictive nature of state preemption. BEAD expects states to give their local governments more agency in determining how broadband access and expansion plans will be charted (Schwartzbach, 2022). One might think states would have no choice but to repeal municipal broadband restrictions. However, not all state broadband programs acknowledge the conflict with BEAD, and those which do, may lack the authority to repeal state legislation (Ali et al., 2024).

The cases in this chapter help us understand how local governments are utilizing ARPA funds to expand broadband connectivity despite constraints from state policy. We profile Palm Beach County, Florida; York County, Pennsylvania; and Shenandoah County, Virginia (for more detail, see Chen et al., 2022). We conducted interviews with local government officials, non-profit organizations, and Internet service providers in October and November 2022. Interviewees included: in Palm Beach, FL: Dr. Adam Miller, Councilman of the Village of Royal Palm Beach and former Initiative Lead from the Performance Accountability Division of the School District of Palm Beach County (SDPBC); and Mike Butler, Network Services Director of Palm Beach County; in York County, PA: Silas Chambers, Vice President of Economic and Community Development at York County Economic Alliance (YCEA); Brian Snider, CEO of Halo Fiber and former Chief Executive Officer of Lit Communities; in Shenandoah County, VA: Jenna French, Director of Tourism and Community Development; Doug Culler, Director of Network Services at Shenandoah County Public Schools.

Across the three cases, we found that state preemption or restrictive regulations were not the primary barriers; instead, we found funding and political will and overcoming incumbent ISP resistance were the major challenges to local broadband deployment and expansion. Innovative partnerships help resolve these challenges while providing localities with benefits such as increased funding, outside management expertise, and protection against state preemption. Partnerships are part of our multi-level governance framework, outlined in Chapter 1, and an element in the resilience framework outlined in Chapter 4. The crucial role of partnerships, especially with mission-driven ISPs, is highlighted in this chapter.

Palm Beach County, Florida Partners with the School District

Florida is one of the 16 states with preemptions on broadband, but this did not stop Palm Beach County from moving forward. The Digital Inclusion Initiative, carried out by the School District of Palm Beach County (SDPBC), was first promoted in 2006. The goal of the initiative was to grant all students access to reliable Internet access in their homes. Due to the economic downturn and lack of new coalition partners, the project slowed down between 2006 and 2019. "Funding was very important to make that happen," said Dr. Adam Miller, the Initiative Lead at SDPBC, who admitted that the lack of financial support greatly set back the Digital Inclusion Initiative in the late 2000s. Because of the high cost of broadband infrastructure at that time, the main focus was to provide, refurbish, and replace the devices for families and organizations in need.

In 2020, the COVID-19 crisis prompted the county's public schools to implement online classes, making it even harder for families without connectivity. This time, Palm Beach County received funding from four sources to address the digital divide, specifically focusing on students without access to adequate high-speed Internet connections. These funding sources consisted of the CARES Act, ARPA funding, Title IV Part A from the US Department of Education, and fundraising support by the Education Foundation of Palm Beach County. However, federal restrictions prevent more than one federal funding source from being used in broadband Internet-related programs. As part of the state's Broadband Opportunity Program, Florida also prohibits the use of duplicate funding in areas where federal funds have been awarded.

Preemptive statutes have harmed the ability of Palm Beach County to provide broadband Internet services. Compared to other public utilities or services, municipal broadband networks in Florida are restricted by state preemption with "ad valorem" taxes and red tape (Cooper, 2024). Florida statute 350.81 prohibits governmental entities from exercising their power in any area to require residents to use or subscribe to any communication service of a governmental entity. It also requires governmental entities to hold at least two public hearings, during which local officials must offer a roadmap to profitability within four years. "Florida statute 350.81 prohibits the county from delivering broadband service to the general public in areas not owned by the government (such as parks, libraries, etc.)," said Michael Butler, the Director of Network Services of Palm Beach County.

To address this preemption, the county targeted only students and their families. "By keeping the signal password protected and limiting extender distribution to students via the School District, we are able to work around this limitation," said Mr. Butler, "but it does limit our ability to serve the broader population." The initiative team (including Miller and Butler) also worked with their legislative affairs team to get language placed in 350.81, which will allow an exemption to serve the underprivileged. If approved, the exemption will allow Palm Beach County to continue to close the digital equity gap.

Navigating Funding Restrictions

Despite funding restrictions, the Digital Inclusion Initiative managed to navigate these parameters and utilize various funds for different processes. The county received $15,750,000 from the CARES Act, which ended in December 2020, and $40 million from the ARPA funds. These two funds were mainly used to build fiber optics and the Wi-Fi mesh network so that the coverage of the network across the county can be expanded. In addition, Title IV Part A provided $70,000 to purchase Wi-Fi extenders for students who live within the Wi-Fi mesh network. Due to restrictions from state funding, the Digital Inclusion Initiative chose to fundraise with assistance from the Education Foundation of Palm Beach County. The Education Foundation connects Palm Beach County's public school system, the private sector, and the community.

The CARES and ARPA funding allowed the initiative to make an ambitious blueprint of extending the key areas identified by the heat maps, installing more than 1,000 miles of fiber optics, and supporting approximately 11,000 radios covering 450 square miles. The Education Foundation has also successfully raised $962,939 to purchase a Wi-Fi extender. Up to September 2022, $66 million had been invested in the county-wide project to support approximately 25,000 students. The different foci of the funds meant that their uses do not overlap. With most funding aimed at capital-intensive broadband middle-mile infrastructure and expansion, money from flexible funds and fundraising was used on Wi-Fi connectors and Internet adoption.

Political Will Against Preemption

The political will from different levels in the county allowed the Initiative to move forward. The construction of fiber optics and other middle-mile programs is under Palm Beach County Network Services, while the School District is responsible for the Digital Inclusion Initiative. To ensure the construction plan is aligned with the strategy of the Initiative, SDPBC shares the updated heat map with the county to direct the plan into key unserved areas. The initiative holds meetings every other week to ensure different parties are on the right track.

Collaborations with municipalities also help with the deployment and fundraising processes. SDPBC first made a partnership with the city of Delray Beach to provide thousands of families with Wi-Fi connectors, laptops, and digital training. So far, 13 municipalities have become part of the Initiative, including South Bay and Belle Glade. The partnerships guaranteed the Right-of-Way permitting for the radio poles, where they can deploy Wi-Fi Mesh Extenders for families in the neighborhoods. The initiative team also helped the Education Foundation connect with partnering agencies in their communities and contribute to the fundraising of the Digital Inclusion Initiative. In every newsletter of the Digital Inclusion Initiative, they update the implementation tracker for the public with progress in different municipalities from fiber optic cable deployment, monopole installations, radio, and antenna installation, Wi-Fi network, and Wi-Fi extenders distributions to students. This allows governments to plan the next steps together, and the public to oversee their progress.

Addressing Equity

For the School District of Palm Beach County, students without Internet connections are their primary concern, especially since COVID-19. SDPBC sent out a survey to students and 170,000 families returned the survey saying they did not have adequate Wi-Fi. Realization of the need pushed leaders to take action on equity, even though they did not explicitly speak of equity.

To ensure a fair process, SDPBC regularly updates the heat map to identify underserved households without Internet access. The team targeted neighborhoods with a high concentration of households in the National School Free and Reduced Lunch Program as their priority to deploy broadband infrastructure and Internet equipment. More areas will be served when more funding is available to extend the project.

The county also emphasizes equity in the adoption process. Considering the lack of knowledge in installing Wi-Fi extenders and using the Internet, a one-page simple instruction sheet (in multiple languages) is given to each household. A QR code on the sheet provides video tutorials to the families. Moreover, each school has a member of staff appointed as the Wi-Fi Warrior, and each community has a community navigator who can help each family connect and learn how to use technology better.

For students living in places without fiber cables or poles installed, the Digital Inclusion Initiative has provided about 4,000 free hotspots in the short term with a Comcast Internet Essentials sponsorship, Sprint Hotspot, or T-Mobile. Once the middle-mile infrastructure is ready, Wi-Fi connectors will be distributed to households for long-term use. The flexibility of solutions to address equity at the community level contributes to the success of the Digital Inclusion Initiative of Palm Beach County.

York County, Pennsylvania Partners with Lit Communities

York County is a largely rural county in southern Pennsylvania, where three-quarters of the county does not meet the FCC standard for broadband speed. Before COVID-19, the county "did not have any broadband strategy" and community members' daily lives were encumbered by slow or inoperable Internet. Silas Chambers, Vice President of YCEA, noted the importance of expanding broadband connection in the county, citing that the "entire county is served by [essentially] one provider." When CARES funding arrived, it triggered a collaboration between YCEA and Lit Communities, a mission-driven company that assists communities to develop a business plan to design and finance fiber optic broadband infrastructure, and leverage PPPs to build and operate the network (VETRO, 2022). York County enlisted Lit Communities for a plan to construct a project to expand broadband connection.

One barrier York County and Lit Communities identified was the lack of state guidance on broadband projects. Title 66 of the Pennsylvania General Assembly prohibits municipalities from being broadband service providers unless there are no service providers or private service providers willing to provide service within "14 months" (66 Pa. Cons. Stat. Ann. § 3014(h)). However, York County was not interested in being the ISP. Instead, they only wished to improve the quality of residents' broadband with this project. They are partnering with Lit Fiber – York, a subsidiary of Lit Communities, as their ISP.

The Power of a Public-Private Partnership

The first stage, funded by the CARES Act, was built in 2020 along the Heritage Rail Trail, a local park. Public ownership secured the rights of way, and 16 miles of county-owned middle-mile fiber was built. This initial project served as proof of concept for the county and Lit partnership. The 16 miles built were integrated into the new broadband expansion project.

Currently, York County and Lit Communities are using $25 million of the county's ARPA budget to deploy middle- and last-mile fiber in this expansion project. Lit has made a plan to build seven rings of fiber access for the middle mile. $20 million from ARPA funds will be spent on the deployment of three rings and 144 miles of fiber in identified underserved areas, mainly in the rural southern part of York County. This will be the first phase of the project; additional capital is needed for the rest of the plan. The last-mile part of the project is focused on fiber-to-the-premises (FTTP) in the urban areas, specifically the City of York and the Borough of Hanover. The FTTP model costs $5 million to improve the quality of Wi-Fi connection and the affordability of services in these city areas. The last mile is projected to finish at the same time as the middle mile. As of 2023, the project was still in the contracting stage, getting the engineers and other contractors to sign on and begin construction. The project is expected to be completed in one to two years.

Without access to these federal funding sources, the project would not have started. Likewise, lack of funding is hindering the completion of the full project; the rest is contingent upon gaining additional capital. However, revenue from leasing middle-mile fiber access to private providers could be a potential source to fund the expansion of all seven rings, along with future public-private partnerships with ISPs and potential state grants or other funding opportunities.

Pushback from Incumbent ISPs

Incumbent providers were an obstacle. YCEA Vice President Silas Chambers explained that local private providers are against any municipality's involvement in broadband. Incumbent providers usually shut down any county interference in broadband by bolstering the preemption law and threatening that what the municipality is doing is *"illegal"* or *"unethical,"* which usually stops municipalities from continuing, says Chambers. He notes that York has *"moved so fast"* on the project that this kind of pushback couldn't work.

York County self-identifies as pro-business. "If we thought that the private sector could do this without us, then we would get out of the way," says Chambers, "but the clear evidence is that they can't or [really] they won't." Incumbent Internet providers weren't properly serving the community. Many residents had Internet packages which didn't provide the speeds advertised and were expensive. The county's project hoped to improve the affordability and quality of Internet York County residents would receive. The county initially conducted a "demand aggregation study" where residents could participate in a survey from their homes, to record speed tests, their current Internet bill, the initial package they bought (what Mbps/Gbps service and for what price), etc. This survey helped the county assess the gap in what the incumbent providers said they were offering and what York County residents were actually receiving.

"Many people are overpaying for the Internet and not getting the speeds that are advertised," says Chambers. York County and Lit Communities designed their broadband project to explicitly address equity concerns. They wanted to improve the quality, accessibility, and affordability of broadband service. They identified underserved and low-income areas across the southern rural part of the county. "It's all about partnerships and doing it the right way," said Brian Snider, then CEO of Lit Communities. The operational structure of the broadband deployment will be that the county will own the fiber, and Lit Communities will provide affordable service through their local ISP, York Fiber.

Partnership with ISP York Fiber LLC

Specifically, the county's ownership of dark fiber (middle mile) will allow them to sell connections to private ISPs, including York Fiber. For example, once connected to fiber, York Fiber will connect the customer, and pay York County for dark fiber access. This will be a revenue-sharing model between York Fiber LLC and York County, which will help the county reinvest in the system. "It's going to be priced for life," says Snider, who detailed the role York Fiber will play. For affordability, York Fiber LLC plans to offer permanent gig packages under $100/mo for speeds which exceed the FCC 25/3 Mbps models. They also plan to include ACP packages, including 100/100 or 250/250 symmetrical depending on the area and what is physically feasible. These packages will be affordable due to wholesale fees. The price of the Internet can be offset by the sale of other products like security systems to companies and businesses being served. In fact, this plan bolsters connection with private ISPs because they can now attach to already existing fiber in areas that previously were out of network and thus, expand their customer base.

The main goal of this project is to sustain the affordability and quality of broadband services. Once the fiber is laid, York Fiber LLC is poised to offer digital literacy and hire locally for long-term maintenance. Both Lit Communities and York County emphasized they did not just want to lay fiber down, but to help residents utilize and benefit from this project. York Fiber plans to train and hire locally for the operation and maintenance of the last-mile connections, customer service call centers, and tech crews. York Fiber will not only connect homes to the Internet but also provide guidance and assistance to users on how to use their new broadband service.

Shenandoah County, Virginia Partners with Telecommunications Company, Shentel

In Virginia, four in ten residents (43.7%) are unable to purchase a fiber Internet plan (BroadbandNow, 2022). Shenandoah County is very rural with population centers in its six towns. It took a coordinated effort between county organizations, the County Chamber of Commerce, and its towns to promote fiber optic infrastructure deployment. Shenandoah County is motivated to deliver high-speed broadband to its residents, whether they live in the densely populated areas or in more remote areas.

From the COVID-19 pandemic to early 2022, DSL, cable, and wireless Internet initiatives were the county's main approaches to expanding Internet access to customers. When Shentel and Shenandoah County accessed the Virginia Technology Initiative (VATI) grant, they were required to focus on fiber optic technology.

The Shenandoah Telecommunications Company (Shentel) is a publicly traded company that seeks to expand broadband Internet access to the county's unserved and underserved areas. For close to 120 years, Shentel has been a vital partner to county public schools, government, and local businesses. Shenandoah County's current comprehensive plan, with Shentel's technical expertise, will enable its six towns to embrace the speed and increased connectivity that fiber offers. Broadband expansion in Shenandoah involves connecting service area customers to FTTH or FTTP delivery models. Previously, 99% of the county used DSL Internet. According to the Virginia Department of Housing and Community Development, Shenandoah County's current project will reach 4,090 residences, 42 businesses, three community anchors, and four non-residential customers with FTTH service. Shentel intends to engage in new residential FTTP expansion by connecting Shenandoah customers to Shentel's expansive multi-state fiber network that includes "...points located in Ashburn, VA and Atlanta, GA (Shenandoah County Public Schools/ Shenandoah County of Virginia, 2021)." This intended FTTP expansion considers the future need for a fast and scalable network that maintains network speed (10 Gbps) despite increased customer demand. Increased Internet connectivity can be viewed as a catalyst for regional economic growth, which is expected to ease affordability constraints that customers face over time.

Preemption, Partnerships, and Funding

Virginia state laws allow municipalities to build their own broadband networks and provide retail services to residents, but they are prohibited from subsidizing services and charging rates that are lower than those of existing service providers for comparable services. The Commonwealth of Virginia lays out explicit aspects that its local governments must adhere to, or from which they are restricted. These regulations reflect a form of preemption, limiting the ability of municipalities to fully control their broadband initiatives and compete directly with private providers.

Collaboration is important in a state preemption environment. Shentel and the Shenandoah Valley Electric Cooperative have signed an agreement to collaborate on a strategy to extend broadband to unserved areas of Shenandoah County, and, in exchange for right-of-way easements, Shentel has formalized a "resource sharing agreement" with the Virginia Department of Transport (VDOT). Shenandoah County is in similar talks with Dominion Energy (DE), as the electric company plans to install middle-mile fiber to help ISPs like Shentel reach its most rural customers.

Due to Shentel's extensive knowledge of the county, meeting the state preemptive feasibility study requirement for the FTTH and FTTP Internet solutions was straightforward. This is important because a number of unserved locations in and around the county almost border another ISP's territory. A positive feature of Shenandoah County's broadband expansion is that the feasibility studies enable project developers to minimize fiber service overlap with existing wireless services around target areas. Shentel's FTTH model minimizes the chances of overlap with other existing Internet providers, as there was no use of the federal Rural Digital Opportunity Fund (RDOF) within and around the broadband expansion areas that Shentel targets for its fiber network.

Jenna French, Shenandoah's Director of Tourism and Economic Development, noted that a private company had approached the county in an attempt to gain a share of the county's proposed broadband expansion, but given the county's ongoing strategic planning with Shentel, a tight turn-around for submitting the state grant, and higher forecasted expense costs from the competitor, sticking with the incumbent proved to be the most feasible choice.

Doug Culler, Director of Network Services at Shenandoah County Public Schools, noted the advantage of being able to combine ARPA funds with the Virginia Telecommunications Initiative (VATI) grant to allow Shenandoah County to have a seamless process of planning and budgeting. A balanced funding mix was the most effective way to achieve the broadband expansion goal set in the most recent County comprehensive plan, and the availability of state, federal, and Shentel funding closed the funding gap.

Broadband deployment funding for Shenandoah County's current project comes from the state's Department of Community and Housing Development's VATI, Shentel, and ARPA. Shenandoah County with Shentel as a partner applicant, received $12.1 million in grant funding from VATI, and plans to leverage $20.7 million from other sources; $17 million from Shentel and $3.7 million from the County's ARPA grant funding. There is potential for a sustainable, well-funded and maintained local broadband network that earns public trust.

Lesson – The Power of New Partnerships

Preemption is a limiting factor for local initiatives because municipalities must abide by their state's respective preemption laws. In addition, municipalities are often hesitant to embark on public broadband projects because they lack the funding and technical expertise to do so. PPPs can help address both these challenges. Involving a private partner, which can construct, maintain, and operate a broadband network, can be more efficient and effective than direct municipal control. When localities work with a private partner, they can overcome the barrier of preemption. In both Brownsville, TX and York County PA the partner was Lit Communities, a mission-driven company that assists communities with broadband infrastructure planning, financing, buildout, and operation. Both communities faced opposition from incumbent ISPs. In Shenandoah County the partner was the regional telephone company and the electric provider. These are the kind of non-traditional providers identified by BEAD and already being funded in some states, as we showed in Chapter 3.

Political will, the commitment of community leaders and stakeholders to take action on a particular matter (Post et al., 2010), was key. In each of these cases, broadband expansion was not planned and carried out until the municipality developed the political will to explore and implement broadband and/or funding initiatives. The election of new municipal leaders focused on broadband, and community realization of broadband's importance following the devastating effect of poor Internet connection during the COVID-19 pandemic, became key motivators that led to quick action on broadband issues.

By far the largest roadblock to implementing the projects across these cases was a lack of funding. ARPA funding was the main catalyst for these cities and counties to start planning and implementing these broadband projects. Without public funding these projects would not have happened.

Opposition from incumbent ISPs was a hurdle, but one that could be overcome with thorough data collection. Incumbent ISPs strongly resisted municipal-led broadband projects in Brownsville, Texas (Chapter 4) and York County, PA. In Brownsville, incumbent, large ISPs, despite failing to participate in the PPP bidding process or to promise service expansions, made various attempts to stall and derail the project. Incumbents paid for advertising campaigns to boast their services and enlisted the local chapter of Council for Citizens Against Government Waste to file Freedom of Information Act requests demanding the release of business information (see Chapter 4 for more detail). In York County, PA, incumbent providers tried to derail the project by arguing that the county could not provide the service. In both cases, municipalities were backed by survey data and clear legal understanding, so the incumbent challenges could be countered. Municipal broadband efforts must have sufficient data to justify the need for broadband improvements.

Finding a supportive partner locally or from outside the region can provide municipalities with many benefits. In our case studies, municipalities found private mission-driven ISPs willing to provide additional funding for local projects and expertise. In Shenandoah, the electric cooperative is helping address access to devices and upgrade electrical infrastructure. In Brownsville, TX, the private partner, Lit Communities was willing to match three times the city's planned investment, significantly increasing the reach of the broadband project. Involving private sector expertise can help the city reduce maintenance and operation costs while saving the local government from having to build a customer support network from scratch. In both Brownsville and York County, the private partner is committed to training and maintaining a local workforce to tackle broadband problems. However, Lit Communities is under new ownership, which may not retain the same vision to serve people who were left behind by the big companies.

One common theme seen across all the case studies was that the localized effort helped address the needs of different community groups. Many communities focused on distributing ARPA funds in a way that would bring broadband to underserved populations. Brownsville is committed to universal coverage within its boundaries, while York, Palm Beach, and Shenandoah are prioritizing unserved areas, underserved areas, or low-income populations. Some locations are also expanding beyond simply making broadband accessible by providing device and technological support.

Preemption is of growing concern for local governments (Briffault et al., 2023). But recent federal policy under the American Rescue Plan is helping to reduce state preemption (Xu and Warner, 2024). In this chapter, we have shown how state preemption of municipal broadband does not have to be a barrier to broadband expansion. We have profiled how local municipalities have developed innovative solutions to maintain some degree of local control, improve service, and still be in compliance with state regulations.

While preemption can limit broadband deployment, lack of money and incumbent ISP opposition are greater issues. These issues can be addressed by innovative partnerships. When municipalities build political will and assume a pragmatic, problem-oriented approach, they can move forward.

Municipalities can find partners in their broadband efforts. Incumbent local providers can be a good start, but other mission-driven entities, like Lit Communities, or electric and telephone cooperatives can provide capital and technical support. These partners can help reduce the municipality's capital cost, provide maintenance and customer service, and amplify the impact of the municipality's investment.

References

Ali, C. (2021) *Farm-Fresh Broadband: The Politics of Rural Connectivity*. The MIT Press. https://doi.org/10.7551/mitpress/12822.001.0001

Ali, C., Berman, D. E., Forde, S. L., Meinrath, S. and Pickard, V. (2024, May 16). The Bad Business of BEAD. Benton Institute for Broadband & Society. https://www.benton.org/blog/bad-business-bead

Block, B. (2022, March 30) How Comcast and Other Telecoms Scuttle Rural WA Broadband Efforts. Cascade Public Media. https://crosscut.com/news/2022/03/how-comcast-and-other-telecoms-scuttle-rural-wa-broadband-efforts

Bravo, N., Warner, M. E. and Aldag A. (2020). Grabbing Market Share, Taming Rogue Cities and Crippling Counties: Views from the Field on State Preemption of Local Authority, Dept. of City and Regional Planning, Cornell University, Ithaca, NY. https://labs.aap.cornell.edu/sites/aap-labs/files/2022-08/Bravo%20et.al_2020.pdf

Briffault, R., Camacho, J., Davidson, N., Nelson, S., Roy, M. and Su, R. (2023). The State Strikes Back: Death Star 2.0 Preemption. Local Solutions Support Center. https://static1.squarespace.com/static/5ce4377caeb1ce00013a02fd/t/6551bf1bbd4f391e4e22f513/1699856159185/LSSC-WhitePaper-DeathStar20.pdf

BroadbandNow. (2022). Internet Access in Virginia: Stats & Figures. BroadbandNow. https://broadbandnow.com/Virginia

Chamberlain, K. (2023, May 9) 2020 Report: Municipal Broadband Is Roadblocked Or Outlawed In 22 States. BroadbandNow. https://broadbandnow.com/report/municipal-broadband-roadblocks-2020

Chen, M., Guo, E., Maduakolam, D., Shen, D., Bowman Brady, J., Olafare Olagbaju, S. (2022). Access, Range, Partnerships, Adoption: Case Studies of U.S. Broadband Expansion Projects. Department of City and Regional Planning, Cornell University. https://labs.aap.cornell.edu/local-government-restructuring-lab/student-work/access-range-partnerships-adoption-case-studies-of-us-broadband-expansion-projects

Common Cause. (2021). Broadband Gatekeepers: How ISP Lobbying and Political Influence Shapes the Digital Divide. https://www.commoncause.org/wp-content/uploads/2021/07/CCBroadbandGatekeepers_WEB1.pdf

Cooper, T. (2024, September 17). Municipal Broadband Remains Roadblocked in 16 States. BroadbandNow. https://broadbandnow.com/report/municipal-broadband-roadblocks

Frug, G. E. (2001). *City making: Building communities without building walls.* Princeton University Press.

Gonzalez, L. (2018, January 30). Community Broadband Pushed Out of Pinetops, N.C. Institute for Local Self Reliance. https://ilsr.org/articles/community-broadband-pushed-out-of-pinetops-n-c/

Goodman, C. B., Hatch, M. E. and McDonald III., B. D. (2021). State Preemption of Local Laws: Origins and Modern Trends. *Perspectives on Public Management and Governance*, 4(2), 146–158. https://doi.org/10.1093/ppmgov/gvaa018

Handgraaf, B. (2017, June 30) Pinetops Broadband Gets New Life. *The Wilson Times*. https://restorationnewsmedia.com/articles/local-news/pinetops-broadband-gets-new-life/

Holmes, A. and Al Idrus, A. (2014). Chattanooga Asks FCC for Help in Spreading Broadband. The Center for Public Integrity. https://publicintegrity.org/inequality-poverty-opportunity/chattanooga-asks-fcc-for-help-in-spreading-broadband/

Kim, Y. and Warner, M.E. (2018). Shrinking Local Autonomy: Corporate Coalitions and the Subnational State, *Cambridge Journal of Regions, Economy and Society*, 11(3), 427-441. DOI: 10.1093/cjres/rsy020

Mitchell, C. (2012). Broadband At the Speed of Light: How Three Communities Built Next-Generation Networks. Institute for Local Self-Reliance. https://ilsr.org/wp-content/uploads/2012/04/muni-bb-speed-light.pdf

Murawski, J. (2014, August 28) Wilson, N.C. Asks FCC to Override State's Telecom Law. Government Technology. https://www.govtech.com/network/wilson-nc-asks-fcc-to-override-states-telecom-law.html

NLC. (2018). City Rights in an Era of Preemption: A State-by-State Analysis. https://www.nlc.org/wp-content/uploads/2017/02/NLC-SML-Preemption-Report-2017-pages.pdf

Post, L. A., Raile, A. N. W. and Raile, E. D. (2010). Defining Political Will. *Politics & Policy*, 38(4), 653–676. https://doi.org/10.1111/j.1747-1346.2010.00253.x

Schwartzbach, K. (2022). With Billions for Broadband Incoming, How Have State and Local Governments Expanded High-Speed Internet Access? Rockefeller Institute of Government. https://rockinst.org/blog/with-billions-for-broadband-incoming-how-have-state-and-local-governments-expanded-high-speed-Internet-access/

Shenandoah County Public Schools/Shenandoah County of Virginia. (2021). Shenandoah County VATI Broadband Grant Application 2022. https://www.dhcd.virginia.gov/sites/default/files/Docx/vati/FY2022-vati-applications/1-shenandoah-county-vati-broadband-grant-application.pdf

VETRO (2022, February 4). Case Study: Lit Communities – Mission Driven Partnerships Are the Key to a More Connected Country. https://vetrofibermap.com/case-study-lit-communities-mission-driven-partnerships-are-the-key-to-a-more-connected-country/

Wen, C., Xu, Y., Kim, Y. and Warner, M. E. (2020). Starving Counties, Squeezing Cities: Tax and Expenditure Limits in the US. *Journal of Economic Policy Reform*, 23(2), 101–119. https://doi.org/10.1080/17487870.2018.1509711

Xu, Y and Warner, M.E. (2024). Fiscal Federalism, ARPA and the Politics of Repair, *Publius: The Journal of Federalism*, 54(3): 487–510. https://doi.org/10.1093/publius/pjae019

PART 3
Institutional Leadership for Digital Equity

Digital equity requires more than physical infrastructure. It also requires attention to affordability and adoption (devices and training in how to use them). In this section, we explore the role of Indigenous nations, ConnectHomeUSA, and Cooperative Extension in helping communities address the digital divide. Each of these initiatives emphasizes the role of collective action, learning across communities, and providing training via digital navigators to ensure that residents can access and actually use the Internet. By building from cultural assets and promoting inter-generational sharing, these initiatives show how bridging the digital divide can strengthen and integrate communities.

In Chapter 8, we showcase efforts by the Choctaw Nation of Oklahoma and the Central Council of the Tlingit and Haida Indian Tribes of Alaska, which have built public–private partnerships and used innovative technology, financing and education approaches to help their residents bridge the digital divide. In Chapter 9 the role of Cooperative Extension in surveying digital skills needs and designing educational tools is profiled. Digital capital is one element of the community capitals framework. Digital access and adoption is essential for the exchange of information and resources, as well as support for financial well-being, business viability, career development, education, healthcare access, and social/emotional connections. Extension has built a Digital Skills Toolkit to create safe pathways of entry into the digital world.

DOI: 10.4324/9781003619208-10

8

INDIGENOUS LEADERSHIP IN ADDRESSING THE DIGITAL DIVIDE

Duxixi (Ada) Shen, Mildred E. Warner, and Natassia A. Bravo

Indigenous communities have worked to close the digital divide and, in doing so, demonstrate the benefit of collective action and cross-community learning in addressing broadband adoption. We expand upon our multi-level governance framework from Chapter 1, which emphasizes the intersection between federal, state, and local policy, and in this chapter showcase the unique role played by Indigenous governments as sovereign nations. While Chapter 6 included examples of Indigenous communities in Colorado and Maine as partners in broader regional efforts to bridge the digital divide, in this chapter we profile two cases where Indigenous communities led the effort. We note how Indigenous communities can pursue a more collective approach to engage community resources, beyond that outlined in our resilience framework in Chapter 4, and promote strategic action to foster community resilience in broadband delivery.

Internet access enables the creation and exchange of information and knowledge. In her 2017 book *Network Sovereignty*, Marisa Duarte sheds light on how Internet access, specifically, is fundamental to sovereign nations' exercise of their cultural sovereignty and economic self-determination. However, several barriers stand in the way – the uneven distribution of broadband infrastructure, the costs of home Internet subscriptions and devices, limited data on Internet usage and digital literacy in tribal communities, difficulty in accessing federal broadband funding, and overlaps between tribal and federal/state authority. Geographic, political, economic, historical, and cultural considerations shape how broadband infrastructure is planned and deployed by sovereign nations (Duarte, 2017).

Federal and state support can expand or constrain the ability of tribal communities to design and implement broadband networks. The Tribal Broadband Connectivity Grant Program provides funding to tribal governments, educational

DOI: 10.4324/9781003619208-11

institutions and organizations for infrastructure, affordability programs, and digital inclusion efforts. The now-defunct Affordable Connectivity Program (ACP) assisted tribal households with up to $75/household subsidies for monthly Internet service charges. Tribal communities are also eligible for infrastructure grants from several other programs run by the FCC, NTIA, and USDA (see Chapter 1, Appendix Table 1.1).

Like their non-tribal counterparts, project design and implementation in tribal communities are impacted by regulatory, geographic, financial, demographic, and technological constraints. The process is also informed by historical, environmental, political, and cultural concerns. For example, network planning and ownership have different implications for tribal governments. By leading the design and implementation of major infrastructure projects, tribal communities can minimize the physical impact on the landscape. By retaining ownership, tribal communities can ensure that everyone is served (Duarte, 2017). Some tribal communities are more remote than others, and thus will need to address different geographic challenges and technology needs. In Alaska, the Haida Tribal Council was able to expand wireless access through broadcast Internet in the remote Wrangell Islands. Similarly, in Oklahoma, the Choctaw Nation worked with multiple providers to offer both high-capacity fiber optic Internet and flexible wireless service.

To tackle the digital divide in Indigenous communities, Duarte et al. (2021) introduce the Full-Circle Framework which emphasizes guiding broadband deployment through four key phases: understanding the sociotechnical landscape, strategizing for future infrastructure, augmenting that infrastructure, and reflecting on its impacts. The geographic and technical challenges are especially difficult in remote Indigenous communities. Duarte et al. (2021) highlight the importance of cultural sensitivity, embracing a collaborative approach, addressing digital inequities, and upholding the principles of sovereignty and self-determination in broadband planning and deployment in Indigenous nations – principles that are reflected in our case studies. They also emphasize the importance of reciprocal relationships between researchers and Indigenous communities.

Other cases in this book have focused more on access. But the cases here give broader attention to adoption and affordability as well. The Digital Inclusion Ecosystem framework, developed by NDIA (2021), emphasizes the existence of programs and policies to address all aspects of the digital divide – not just access to broadband service, but the need for technical support in digital literacy and device ownership – typically provided by digital navigators. Collaboration in the community must include policymakers, advocates, and Internet Service Providers (ISPs) along with community leaders (Figure 8.1) – and in Indigenous communities, inter-generational approaches have a special role to play.

Collective action is a key theme observed in these cases. Collective action involves more than collaborative efforts among diverse stakeholders to achieve

Affordable and Subsidized
Broadband Service Options

Entities Providing
Local Digital Inclusion Services

Affordable and Subsidized
Device Ownership
Programs

Policymakers

Multilingual
Digital Literacy and
Digital Skill Trainings

DIGITAL
INCLUSION
ECOSYSTEM

Existence of
Programs and
Policies
Addressing All Aspects
of the Digital Divide

Collaboration in
community
to Co-create
Solutions in
Partnership

Advocates

Hardware And Software
Technical Support

Social Service
Providers

Digital Navigation Services
to Guide to the Above
Services

Community Leaders

FIGURE 8.1. Digital inclusion ecosystem.

Source: Author, based on the definition of Digital Inclusion Ecosystem by NDIA (2021).

common goals. Here it involves coordinated initiatives by Indigenous government agencies, housing authorities, non-profit organizations, and private industry stakeholders to bridge the digital gap. Unlike other rural communities, Indigenous communities have governmental authority, which enables a broader collective action approach. They face additional challenges due to rural, remote geography, poverty, and discrimination. However, they can also build on cultural patterns of deep community resilience which enable them to address the digital divide in creative ways.

Our cases, selected from different geographical regions and demographic contexts, find that the influence and collaboration built by cross-regional organizations help reduce some of the barriers to implementing broadband services. Regional organizations like the Central Council of the Tlingit and Haida Indian Tribes of Alaska (CCTHITA) have more power than individuals to organize and implement projects, because they build scale while establishing and maintaining credibility in their communities. Similarly, the Choctaw Nation covers more than ten counties in Oklahoma so it can reach a scale broader than a single community. The federal ConnectHomeUSA program by HUD was key in pulling resources to provide technical assistance and help communities learn from each other.

Launched as "ConnectHome," the program started with just 28 pilot communities in 2015. Two years later, HUD partnered with EveryoneOn, a national non-profit aiming to provide under-resourced communities with adoption resources, to expand the program under a new name "ConnectHomeUSA." The program focuses on narrowing the digital divide by helping local housing agencies across the U.S. connect with nationwide ISPs, non-profits, and the private sector. Every month, public housing authorities (PHAs) under the ConnectHomeUSA program have a call together to ask questions, share their experience, and talk about success. This creates a hub for learning and collaboration for ConnectHomeUSA teams across the country. The Housing Authority of the Choctaw Nation (HACNO) collaborated with ConnectHome, as have housing authorities in cities, such as Seattle and Jersey City (Shen et al., 2024). In addition to addressing the physical infrastructure challenges of Internet access, a housing authority can provide community computer labs and digital navigators to residents for practice, training, outreach, and learning from each other (Turner Lee, 2024; Shen et al., 2024). Financial constraints are a big concern due to the lack of funding allocated to PHAs in ConnectHomeUSA communities (The Council of Large Public Housing Authorities, 2022). The Affordability Connectivity Program has been a key resource to subsidize residents in communities across the US, but with the ACP's loss of funding in 2024, affordability will become even more of a challenge in the future.

The case studies in this chapter are based on a review of publicly accessible documents as well as expert interviews conducted with key players in each community. We interviewed Chris Cropley, Network Architect, CCTHITA, and Luke Johnson, Wireless Systems Architect, CCTHITA. In the Choctaw Nation of Oklahoma, we interviewed Josh A. Raper, Manager of the Connect Home Department for the Housing Authority. The key themes observed across the cases – partnerships in access, funding strategies for affordability, and Digital Inclusion Services in adoption – all stem from collaborative efforts. Through an examination of these common threads, we gain valuable insights into the diverse strategies employed to promote connectivity in tribal communities.

Wrangell, Alaska: Political Will and Collective Action

The driving motivation behind the CCTHITA's broadband initiative was bringing access to underserved residents. Wrangell was selected because it presented an ideal setting for their pilot project.

> "We're starting in Wrangell," Chris Cropley said, a Network Architect at CCTHITA, "It's perfect – it's the Goldilocks, as they say. It's got everything going for it. We've got a lot of people there, relatively, we own the 2.5 [GHz spectrum] there, and we have funding for it." (Smiley, 2021)

Wrangell is located within the Alexander Archipelago, which has a chain of islands that make up much of southeastern Alaska. The Wrangell Islands span the size of Florida but have a total population of only 60,000 residents. The geographic characteristics present a unique challenge for local broadband deployment. According to 2022 US Census data, about 21.8% of Wrangell residents are American Indian or Alaska Native. While the City of Wrangell is fairly dense, smaller towns located on various islands create challenges in terms of building out the necessary infrastructure, logistics of getting people and equipment to each of the islands via ferry, and completing all this work during warmer months.

The Tlingit and Haida Tribal Council had the political will to address the issue of access. The Council took swift action with the FCC's provision of broadband licenses and ARPA funds in order to close the digital divide. While their broadband initiative started before ARPA funds were made available in 2021, the Tlingit and Haida Tribes plan to use the $15 million in ARPA funds to build on the 2.5 GHz of broadband spectrum and bring 4G, 100 Mbps symmetrical wireless connectivity to roughly 10,000 Wrangell City residents (Oxendine, 2022). The initial program started in December 2021 after the Tlingit and Haida Indian Tribes of Alaska were awarded a broadband license by the FCC to improve connectivity. With this license, the Tribe is granted exclusive use of a mid-band broadband spectrum in the Wrangell Islands (Smiley, 2021). The 2.5 GHz spectrum is essentially owned by the Tribe and its broadband service, called Tidal Network, seeks to provide home Internet directly to homes and businesses that do not have existing reliable Internet access.

Once the service is in place, the Tribe will have to "defend" it, which requires them to offer Internet service to 80% of the area's population in the first two years, and full coverage after five years. With a "public-focused effort rather than a competitive, revenue-driven business," the CCTHITA's fixed wireless Internet service, Tidal Networks, has created unified and successful business relationships under the larger mission of strengthening, preserving, and sharing Native cultures and communities with the world (Oxendine, 2022). The link between technological innovation and supporting Indigenous culture is a key priority.

Tidal Networks' solution for the remote, island communities is to build 120-foot-tall towers that will be able to broadcast wireless Internet directly to homes and businesses. They have also leased existing towers and are using cell on wheels (COWs) to provide Internet service from the towers. Because their focus is underserved groups, their emphasis is not on replacing incumbent providers, but to provide access for all regardless of profit. They plan to allow other providers to put up their equipment on the towers. Cropley emphasizes that the public service is "looking at finding the most people with the least Internet or no Internet, ideally, and providing them with Internet where they didn't have before, where they had to use a satellite" (Smiley, 2021).

ARPA funding was instrumental to the success of providing accessible Internet to at least 80% of Wrangell residents by 2023 and full coverage by 2026.

Cropley states, "We don't have a population density problem; we have a middle-mile problem." Without funding, the Tribal Council would not be able to build out the necessary infrastructure required to access their exclusive mid-band broadband spectrum. Additionally, they would not be able to expand their initiative to more rural areas of the Wrangell Islands, as well as develop a 4G wireless solution for the greater Wrangell area. Funding has allowed the Tribal Council to address Internet accessibility issues beyond just the deployment of broadband.

Tidal Network's broadband initiative was explicit in their focus on underserved groups. Citing the eight demographic categories outlined by the Digital Equity Act of 2021, their services seek to target

> individuals living in households with incomes at or below 150% of the poverty line, individuals 60 years of age or older, veterans, individuals living with one or more disabilities, individuals with barriers to the English language, members of racial and ethnic minority groups, individuals residing in rural areas, and individuals incarcerated in a non-federal correctional facility (King et al., 2022).

With this initiative, the Tlingit and Haida Tribes hope to close the digital divide while using the Internet to preserve and share Native culture, art, and language.

Collective Action in Partnerships

The project's success can be attributed to the strong collective action formed through partnerships. It is especially challenging for the CCTHITA to seek out collaborators for last-mile services. "The providers have shareholders to be responsible for, and their job is to make money. Our job is to provide Internet for people," Cropley explained. "It's a little bit different of a business model or paradigm, so it's been hard to find partners, vendors, and consultants that align with that" (Oxendine, 2022).

In 2021, Cropley and a colleague met coreNOC Inc. at the Wireless Internet Service Providers Association event, where they discovered that coreNOC specializes in building rural networks for tribal communities. The arrangement coreNOC developed and procured in Wrangell was described as a "knowledge transfer" opportunity by Johnie Johnson, CEO of coreNOC Inc.:

> "That's what's been most attractive, I think. We help tribes and clients understand how to install, how to service and how to monitor the technology," Johnson said. "We're able to help them select the technology, to deploy the technology, help them commission, and then bring that to market. At the end of the day, we have a full turnkey solution we provide for the tribes." (Oxendine, 2022)

In addition to local efforts for middle-mile deployment, collective action has been instrumental in broadband adoption in Indigenous communities in Wrangell, Alaska. The Wrangell case illustrates the use of a range of technologies (fixed broadcast, 4G) as a potential pathway for tribal communities facing challenging geographic features to build broadband access.

Digital literacy and technical support are the other half of the digital inclusion ecosystem (see Figure 8.1). With their commitment to being a full-service provider solution, the Tribal Council decided to provide support for digital literacy and other network solutions. The next big step for the Wrangell program is addressing digital literacy. Notes Cropley:

> Internet's only half the battle, it's teaching people how to use it, making sure they're using it correctly. We're going to be providing in-home routers that do Wi-Fi optimization and prioritization, so that we're able to facilitate working from home, school from home. We're not just dropping Internet off on your roof and saying 'Good luck.' (Smiley, 2021)

The Choctaw Nation of Oklahoma: Addressing Challenges in Connectivity

The Choctaw Nation of Oklahoma is the third largest federally recognized Tribe in the United States, and located in the southeastern corner of Oklahoma. As a tribal nation, the challenge of having high-speed Internet connections is more acute. According to the Federal Communications Commission (FCC, 2021), 20.9% of the residents in tribal areas lie outside the coverage of fixed terrestrial 25/3 Mbps Internet connection in 2019, compared to 17.3% in rural areas and 1.2% in urban areas. We can be sure these figures are undercounts.

The ConnectHome project in the Choctaw Nation is aimed at closing the digital divide in public and assisted housing. It was launched with the support of the federal government. In 2015, then-President Obama visited the Choctaw Nation to announce the nationwide ConnectHomeUSA program (The White House, Office of the Press Secretary, 2015). Twenty-eight local PHAs across the US, including the Housing Authority of the Choctaw Nation of Oklahoma (HACNO), were selected to join forces in addressing digital equity in their communities.

The Choctaw Nation housing authority has a variety of housing programs for tribal members, including Affordable Rental, Independent Elder Housing, and the Lease to Purchase program. Unlike the properties owned or managed by most housing authorities in the US, many HACNO developments are in remote areas without ISP coverage (HUD, n.d.). Therefore, many households in the Choctaw Nation could not access Internet services from their homes before the ConnectHome program was introduced.

Challenges in Connectivity

The Choctaw Nation faces challenges in connectivity, including limited Internet access, geographic challenges, and limits to tribal governance in the broader region. When HUD first started the ConnectHome project in 2015, the state of Internet access for residents of public housing was poor. The HUD's National Baseline Internet Access Survey showed that 35% of public housing households were under-connected for high-speed Internet access with computers, laptops, or tablets in 2016, and 31% had no Internet access at home (HUD, 2018a). For the Choctaw Nation, the situation was worse – "we found out only 3% had at-home Internet (in 2015)," said Fred Logan, the former manager for the Choctaw Connect Home program, at the 2017–2018 ConnectHome Webinar (HUD, 2018b).

"I mean, we struggled because we are a rural area," stated Josh Raper, the current ConnectHome manager. Activities like visiting educational websites or filing health information and job applications were essential actions that required Internet access for Choctaw Nation residents. However, there weren't any ISPs interested in serving the entire geographic area (HUD, 2018a).

One reason why many ISPs were not interested in building infrastructure is that the terrain of the Choctaw Nation is so large that it exceeds some states like New Hampshire or Rhode Island. According to US Census Bureau and Choctaw Nation website, the population density of the Choctaw Nation is 20.89 per mile, compared to the 57.90 for Oklahoma State, and the housing is spread out. Despite the higher speeds and fewer interruptions, laying fiber and establishing Internet service in these areas would be costly. When fiber is not available, HACNO would provide wireless options by cell towers or coax Internet services as alternatives. However, in some places, these alternatives may show poor connectivity results in speed tests due to the considerable distance from infrastructure towers. The vast geography also posed another barrier for housing authority staff who aimed to contact each resident of HACNO housing to facilitate digital adoption.

In addition to the geographic and technical challenges in infrastructure, the distinctive governance system of tribal nations also affected how the funding and policy are structured. The Tribal Council and its city governments are responsible for all the tribal members. However, this means tribal members living outside the Choctaw official 10.5 county area, and non-tribal residents living in the territory, may find it hard to receive services from public-sector actors like HACNO. "We have people that live outside of the 10.5 counties of the tribal reservation," said Raper, "I wish there was a way we could provide them with Internet services."

As of 2023, there were five Choctaw housing programs enrolled in the Connect Home program: three affordable housing programs for tribal members (Affordable Rental I&II and Independent Elder Housing), one HUD-funded affordable housing program open to all the elderly fulfilling the requirements (202 Supportive Elder Housing), and a Lease to Purchase program (LEAP) funded by the Tribe. For Connect Home projects, only the Choctaw 202 Supportive Elder Housing is not limited to tribal members, as these are open-market properties.

Smart Contracts with ISPs

Currently, the Choctaw Nation of Oklahoma pays for all the broadband infrastructure upgrades. HACNO is usually charged one-time fiber installation fees by ISPs, the price of which depends on factors like location, geological characteristics of the land, and the number of units. By linking access with affordability, HACNO is able to incentivize ISPs to waive the construction costs because HACNO agrees to cover the Internet service payments for households by including these in HACNO's annual budget through a multiple-year contract with ISPs. This ensures effective demand for Internet providers' services. The monthly charge for each household is around $45 on average, and is paid by HACNO. The smart negotiation contracts HACNO makes with the ISPs are possible because of its broad collective action approach and its project design linking access and affordability up front.

HACNO has built partnerships with a range of partners: Cherokee Communications, Dobson Fiber, Pine Cellular, and Vyve Broadband for fiber options, and with T-Mobile and Verizon for hotspots and towers. Wireless options like Verizon jetpacks are issued to tenants during the fiber installation period as a temporary solution for Internet connection. Among these partnerships, Cherokee Communications, Pine Cellular, Vyve Broadband, and Verizon were introduced from the ConnectHomeUSA Initiative back in 2015 (The White House, Office of the Press Secretary, 2015). Every year, HACNO also does an open Request for Proposals to all the ISPs.

The tribal nation covers all the Internet subscriptions for the residents of their housing programs. To achieve this, the Tribe has accessed funding from different sources. In the ConnectHome Manager's opinion, one of the biggest advantages of being a tribal nation in terms of delivering broadband connectivity is that HACNO can leverage both local tribal funds and federal grants. The Internet subscriptions in Affordable Rental Programs and Independent Elder Housing are covered by the Native American Housing Assistance and Self-Determination Act (NAHASDA) funding. The Chief and the Tribal Council also strive to reallocate local funds and leverage local programming. Section 202 Elder Housing program uses the program income generated from the site, and the LEAP program gets its Internet plans paid by the tribal nation out of its own budget for the homeowners.

Tribal Nation Priorities Include Affordability and Adoption

"We are a little different because we're tribal. We have a different budget and different standards," said Josh Raper. For example, the ACP Program offers consumers a monthly benefit of up to $30 for broadband services while providing up to $75 per month if they are residents of Tribal lands (USAC, n.d.). When residents are enrolled with Lifeline, they are automatically qualified for the ACP. Although the FCC has cut back on the Lifeline project now, back in 2015, the program nearly covered every dollar for the service subscription of Connect Home residents. This definitely helped Choctaw's Connect Home stand on its feet, especially at the very beginning of the program.

The main funding source HACNO now uses for the housing program is the NAHASDA. It helps promote housing services and maintenance, and ensures better access to private mortgage markets for Indian Tribes and their members. In 2022, HACNO applied for $13,016,928 in grants from the NAHASDA Indian Housing Plan for FY2023, among which $400,000 was spent on the ConnectHome Program for 600 households. In 2023, HACNO submitted an application for $13,439,758 for FY2024, where an estimated $400,000 will be spent on 200 ConnectHome units (The Tribal Council of the Choctaw Nation of Oklahoma, 2021; The Tribal Council of the Choctaw Nation of Oklahoma, 2022; The Tribal Council of the Choctaw Nation of Oklahoma, 2023).

"Choctaw Nation has their own grant department … that goes out looking for grants and they were able to get us one of these to get us started to help pay for all the Internet service," the previous Choctaw ConnectHome manager, Fred Logan, noted in the 2017–2018 ConnectHome training webinar (HUD, 2018b).

Beyond NAHASDA, the Tribal Council has actively explored and applied for different federal grants. The grant applications for housing and broadband connectivity include the Tribal Broadband Connectivity Program Grant, United States Department of Agriculture (USDA) funds, the CARES Act, and ARPA funding.

HACNO is dedicated to the third pillar of broadband connectivity – *adoption*. It contains three main staff in the team: a manager, a training coordinator who conducts individual and group training for the residents and coordinates events with public libraries for digital knowledge, and an IT project coordinator who handles all the technical issues with all the equipment. HACNO has a contract with Azpen Technology for tablets, which are taken to the residents and used for training.

Collaboration in the community is also emphasized by HACNO. In Indigenous communities, cross-generational connections are especially important. They are core to cultural survival, as elders pass on traditions and language to the young. But for digital inclusion, the young have a special role to play in helping seniors overcome the challenges of adoption. In the Choctaw Nation, the

Housing Authority collaborates with the Youth Advisory Board (YAB), a local program that empowers students in grades 8–12 through leadership projects. Digital training with the elderly is one of the volunteer activities they support.

> … sometimes we will have training, say in the middle of the week on a Wednesday evening at an Independent Elderly (program site). And these are high school kids. But we have them sit with the elders and help them set up emails and other things. And it's also good for the youth and the elders to get to know each other, talk to each other, and kind of learn, said Josh Raper.

The students even do games, throw holiday parties, and host karaoke and dancing events with the seniors, bringing all their favorite activities as well as vibrancy into the community. "It's a win-win for everybody... We furnish a meal for everybody there. And it's turned out really well. Both parties are enjoying it quite a bit" says Raper. The Choctaw Housing Authority also works with Oklahoma libraries to get every resident a library card, so that they can have access to the digital literacy programs and all kinds of video courses.

One benefit of the Choctaw Nation joining ConnectHome was the chance to meet partners who could help the nation address affordability and adoption. For example, Best Buy, one of the biggest supporters of EveryoneOn, promised to offer HACNO residents computer training and technical support to gain the academic and economic benefits of broadband access (The White House, Office of the Press Secretary, 2015). Kano is another partner that the Choctaw Connect Home team met in the 2015 Washington ConnectHome meeting (HUD, 2018b). Kano provides training opportunities for students at different levels of education. Like other housing authorities in the ConnectHome programs, HACNO also works with ABCMouse which provides a code for residents so they can receive a one-year free membership for online courses, including coding.

As of 2023, more than 90% of HACNO residents have Internet access, 77% of whom are connected to fiber. By the end of 2024, they estimate that over 600 housing units in the Choctaw Nation will be completed. HACNO is now working with ISPs to get Internet access fiber laid before the tenants move in. The next step for the Housing Authority of the Choctaw Nation is to make its properties 100% connected to the Internet, the higher the proportion of fiber, the better.

> "That is our struggle every day. We have such rural areas in our 10.5 counties that finding an ISP to provide affordable broadband services is a problem in some of the areas," said the current program manager, Raper, "every year we have a goal of adding more units. Some of the areas are going to be harder, the construction costs are going to be a little higher and we have to negotiate that. We might have to wait a little longer on some of those. But yeah, we will still make sure that they have connectivity, one way or another."

Key Lessons

In Indigenous communities, there are multifaceted access, affordability, and adoption challenges to addressing digital equity. These cases demonstrate how local tribal authorities are working on solutions to these challenges. Four themes stand out.

Collective action through partnerships can build access. In both the Wrangell and the Choctaw Nation cases, collective action through partnerships was crucial to expanding broadband access in underserved communities. Wrangell's collaboration with coreNOC Inc. facilitated the technical expertise needed for last-mile connectivity, while the Choctaw Nation worked with multiple providers to implement both fiber and wireless solutions. These partnerships demonstrate the power of collective efforts to overcome challenges and provide vital digital access to rural and tribal areas.

Affordability requires creative approaches. Pursuing external subsidies has proven to be instrumental for tribal nations to address affordability for their residents. Funding streams like the FCC's ACP and the NAHASDA for tribal nations have provided the housing authorities with a chance to connect those who are among the most disconnected. For example, the Choctaw Nation used guaranteed subscription revenue streams to encourage ISPs to shoulder the investment costs of laying fiber themselves. These examples underscore the importance of actively seeking and accessing diverse funding sources to support broadband expansion efforts and bridge the digital divide in underserved communities. The concern going forward is how to maintain affordability given the sunset of the ACP program.

Adoption requires a community-wide approach. Access is not enough. NTIA's Digital Ecosystem emphasizes training and digital navigators to reach residents and ensure they can use Internet technology. The Choctaw Nation gave serious consideration to this with its special adoption team and the creation of intergenerational programs for training and engagement. Nicole Turner Lee (2024) has emphasized the importance of culturally congruent and accessible programs to ensure those least likely to have access can become connected. The Digital Opportunity Compass (Rhinesmith et al., 2023) recognizes how the Internet connects us to health, employment, education, and recreation opportunities. It is these broader goals that make digital inclusion so critical, and why they are emphasized as foundational to community resilience and individual opportunity. These case examples illustrate how digital inclusion is key to sovereignty and self-determination, a point emphasized by Duarte et al. (2021).

Promote lateral learning across communities. These cases also illustrate the need for linking community initiatives across the nation. Several organizations exist to provide technical assistance and promote learning and sharing across communities. One of them is the Local Initiatives Support Corporation (LISC), which provides guidance to communities in identifying connectivity gaps, engaging with providers, expanding adoption and digital literacy, and improving connectivity in affordable housing. For low-income communities, a key partner has been ConnectHome USA, a program of the US Department of Housing and Urban Development. One benefit of joining ConnectHome is that it holds National ConnectHomeUSA Summits, providing an opportunity for community leaders to communicate and connect with digital experts.

Leveraging networks and learning from other authorities are strategies used in our cases. The monthly calls facilitated by HUD offer a unique opportunity for HACNO and other PHAs to collaborate and learn from each other's experiences. Connections through the ConnectHomeUSA program present an opportunity to learn about what other communities are doing, their funding sources, and partnerships. This can help communities design a model for sustaining programs long-term. EveryoneOn also has helped build various partnerships for housing authorities that participate in ConnectHomeUSA programs. These organizations reduce the financial burden either by offering high-standard Internet services at low-cost prices or by providing digital devices or literacy training to local residents.

Overall, building collective action through community partnerships, applying for funding opportunities, and leveraging networks are essential for addressing broadband accessibility and promoting digital inclusion in communities nationwide. It is essential to address not only the availability of infrastructure but also the issues of affordability and adoption, to ensure that once broadband access is provided, it is financially accessible, and people are equipped with the skills and knowledge to effectively use it. The innovators in our case studies tackled digital challenges in their communities by coming up with effective, creative solutions. Their commitment to digital inclusion reminds us that digital inclusion is not only an effort for technological improvement. It is a critical infrastructure that will help ensure Native Americans have the opportunity to participate in their communities and wider society, as a whole.

References

Duarte, M. E. (2017). *Network Sovereignty: Building the Internet across Indian Country.* Seattle: University of Washington.

Duarte, M. E., Vigil-Hayes, M., Zegura, E., Belding, E., Masara, I., & Nevarez, J. C. (2021). As a Squash Plant Grows: Social Textures of Sparse Internet Connectivity in Rural and Tribal Communities. *ACM Trans. Comput.-Hum. Interact.*, *28*(3), 16:1–16:16. https://doi.org/10.1145/3453862

FCC. (2021, January 19). *Fourteenth Broadband Deployment Report.* https://www.fcc.gov/reports-research/reports/broadband-progress-reports/fourteenth-broadbanddeployment-report

HUD. (n.d.). *CNHA: Fiber Networks and Reliable Internet Service Provided to Remote Sites for the First Time.* Retrieved August 26, 2023, from https://www.hudexchange.info/programs/connecthomeusa/case-studies/cnha-fiber-networks-and-reliable-internet-service-provided-to-remote-sites-for-the-first-time

HUD. (2018a). *ConnectHome Initiative: Final Report | HUD USER.* https://www.huduser.gov/portal/sites/default/files/pdf/ConnectHome-Initiative.pdf

HUD. (2018b, March 27). *2017–2018 ConnectHome—Program Management and Sustainability*. https://www.hudexchange.info/trainings/courses/2017-2018-connecthome-program-management-and-sustainability/2172

King, H., Martin, M., McArdle, S., Goldberg, R., & DeSalvo, B. (2022, May 13). *New Digital Equity Act Population Viewer Shows Broadband Access and Demographic Characteristics*. Census.Gov. https://www.census.gov/library/stories/2022/05/mapping-digital-equity-in-every-state.html

NDIA. (2021). *The Words Behind Our Work: The Source for Definitions of Digital Inclusion Terms*. National Digital Inclusion Alliance. https://www.digitalinclusion.org/definitions/

Oxendine, C. (2022, May 9*). Tlingit and Haida Partner with Native-owned Company to Develop New Internet Provider*. Tribal Business News. https://tribalbusinessnews.com/sections/economic-development/13897-tlingit-and-haida-partner-with-native-owned-company-to-develop-new-internet-provider

Rhinesmith, C., Dagg, P. R., Bauer, J. M., Byrum, G., & Schill, A. (2023). *Digital Opportunities Compass: Metrics to Monitor, Evaluate, and Guide Broadband and Digital Equity Policy*. Quello Center, Michigan State Univ. https://quello.msu.edu/wp-content/uploads/2023/02/Digital-Opportunites-Compass-Paper-20220223.pdf

Shen, D., Redmond, E., Bowman Brady, J., & Warner, M. E. 2024. *Assessing the Digital Divide in Affordable Housing: The Power of Collective Action*. Department of City and Regional Planning, Cornell University. https://labs.aap.cornell.edu/node/881

Smiley, S. (2021, December 21). *Tlingit & Haida to Pilot Its New Broadband Internet Service in Wrangell*. KSTK. https://www.kstk.org/2021/12/21/tlingit-haida-to-pilot-its-new-broadband-internet-service-in-wrangell/

The Council of Large Public Housing Authorities. (2022). Retrieved August 27, 2023, from *Connecting Hope: How Public Housing Authorities Bridge the Digital Divide*. CLPHA. https://clpha.org/sites/default/files/CLPHA-Digital%20Equity%20report-digital-final.pdf

The Tribal Council of the Choctaw Nation of Oklahoma. (2021). *CB-84-21: To Approve the Application for the Tribal Broadband Connectivity Program Grant*. https://www.choctawnation.com/wp-content/uploads/2022/03/cb-84-21.pdf

The Tribal Council of the Choctaw Nation of Oklahoma. (2022). *CB-11–23: To Approve Application for the 2022 United States Department of Agriculture Rural eConnectivity (Reconnect) Program*. https://www.choctawnation.com/wp-content/uploads/2022/10/cb-11-23.pdf

The Tribal Council of the Choctaw Nation of Oklahoma. (2023). *CB-85-23: To Approve the Native American Housing Assistance and Self-Determination Act of 1996 Indian Housing Plan for FY2024*. https://www.choctawnation.com/wp-content/uploads/2023/07/cb-85-23.pdf

The White House, Office of the Press Secretary. (2015, July 15). *FACT SHEET: ConnectHome: Coming Together to Ensure Digital Opportunity for All Americans*. Whitehouse.Gov. https://obamawhitehouse.archives.gov/the-press-office/2015/07/15/fact-sheet-connecthome-coming-together-ensure-digital-opportunity-all

Turner Lee, N. (2024). *Digitally Invisible: How the Internet Is Creating the New Underclass*. Washington, DC: Brookings.

USAC. (n.d.). Affordable Connectivity Program. *Universal Service Administrative Company*. Retrieved August 27, 2023, from https://www.usac.org/about/affordable-connectivity-program/

9

OUTREACH AND EDUCATION RESPONSES TO THE DIGITAL DIVIDE

Research and Action through Extension

*John J. Green, Roberto Gallardo,
and Roseanne Scammahorn*

Although digital information and communication technologies and applications provide significant opportunities, they also present challenges. Critical challenges include becoming aware of this changing landscape, successfully steering away from barriers, and pursuing opportunity pathways in this era now known as the "digital age." As our society and economy continue to digitize core functions and interactions, those on the wrong side of the digital divide will be left further behind and unable to fully participate in community life. This requires adaptation, and change is often better understood and navigated with trusted partners; what could be labeled as community empowerment intermediaries. Here is where the Cooperative Extension Service – across its programmatic areas (agriculture and natural resources, family and consumer sciences, 4-H and youth development, and community resource development), local networks, and long history of being a trusted source of information – can and should play supportive roles. The community resource development area has much to contribute, given the priorities of building economically viable communities and promoting civic engagement and community decision making (Beaulieu and Cordes, 2014).

Providing research-based education, training, and capacity building through Land-Grant Universities for more than 100 years, Cooperative Extension's presence in all states and nearly every county in the nation, coupled with its track record as a convener and intermediary, provides it with a competitive advantage and obligation to help individuals, households, and broader communities in pursuit of their digital transformations. The digital divide is a complex and nuanced issue that typically manifests itself across three dimensions: broadband connectivity, digital skills and utilization, and devices.

DOI: 10.4324/9781003619208-12

All three dimensions need to be addressed for communities to adapt to and benefit from opportunities in the digital age. Although Extension efforts have focused on broader community development to encompass digital transformation, relying on self-help and asset-based approaches (Gallardo et al., 2018), its mission, structure, and expertise aligns particularly well with digital skills and utilization, the efforts addressed in this chapter.

This is especially important in the context of unprecedented investments in broadband infrastructure. Infrastructure, while a vital component in the digital gap, carries little weight without local people gaining the digital literacy skills necessary to engage effectively and safely. This in turn drives the value of home adoption higher, helping make these infrastructure investments sustainable.

There is abundant research making the economic case for expanding digital skills. As noted by Ezell (2021), according to research from the Organization for Economic Cooperation and Development (OECD), one-third of working-age Americans had limited digital skills. However, 70% of jobs in 2016 required medium to high digital skills (Muro et al., 2017), an expectation that has surely increased even more rapidly in recent years. Thus, it is critical that digital skills needs are understood, and outreach and educational tools are designed and implemented. This chapter provides an overview of Extension efforts to do so. It starts with an introduction to the Digital Capital Framework followed by the presentation of a survey conducted to inform Extension programming, and a summary of efforts for the organization and implementation of digital skills training to be offered through community empowerment intermediaries.[1]

Digital Capital Framework

Training intermediaries are needed to help communities address the digital divide. Chapter 8 profiled the work of ConnectHome in public housing authorities and the Digital Inclusion Framework of the NTIA. In this chapter we profile the work of Cooperative Extension in communities across the US. The community resilience framework, outlined in Chapter 4 emphasized the importance of networks and local leadership. Here, we link that to theories that span from the household to the community level. The Livelihoods Approach in development studies focuses attention on the material and experiential needs that individuals, households, and communities have, coupled with the assets and capabilities they have or need to meet those needs (Chambers and Conway, 1992; De Haan, 2012; De Haan and Zoomers, 2003). It also examines how power relations shape which pathways are open or closed (Van Dijk, 2011). Assets, deemed forms of "capital" because they can be invested in, accumulated over time, and exchanged for each other, include natural, physical/built, human, social, cultural, financial, and political. As the approach has been applied to ever-changing development contexts and concerns, additional labels have been developed, including the now common Sustainable Livelihoods Approach (SLA).

The Community Capitals Framework (CCF) (Flora et al., 2016) shares many similarities with SLA, but is noted for shifting the focus more to the broader community system level to address collective access to and development of diverse forms of capital (Gutierrez-Montes et al., 2009). Scholars have argued that investments in diverse forms of capital, especially bonding and bridging social capital, can be reinforcing and result in a spiraling up phenomenon (Emery and Flora, 2006), and limited investment or disinvestment can lead to spiraling down. It is important that diverse social relationships and forms of capital are built, especially those bridging across differences and outside of local networks, to avoid being too insulating and exacerbating social norms that prevent innovation (Newman and Dale, 2005). Recent writing on community resilience has connected SLA and CCF with a focus on how they help us to understand the capacity for communities to adapt to broader social, economic, and environmental changes (Nyamwanza, 2012; Cafer et al., 2024).

Although not typically viewed as a shock in the same way as an economic downturn, natural disaster, or a long-term stressor like globalization or climate change, the sweeping society-wide changes in the digital age increasingly require adaptation for people to achieve their livelihoods and improve community quality of life. SLA and CCF both contribute to how we conceptualize the forms of capital at the individual, household, and community levels and how they can be nourished to improve capacity to handle change. However, they have not traditionally accounted for the ways in which people's channels for accessing information, communicating, participating in education and the workforce, and engaging with goods and service providers have changed in this era. The need for digital access and skills has become so ubiquitous there is a new form of capital to consider – digital capital. We must attend to whether different groups and places are digitally included in these new aspects of social and economic life (Gallardo et al., 2021; Ragnedda, 2018). Just as social capital has been conceptualized as being essential for exchange between the other forms of capital, digital capital is necessary for how we interact and exchange information and resources. Connecting to SLA and CCF, digital inclusion can be seen as critical for people to know about, pursue, and ultimately achieve diverse livelihood pathways (Green, 2024).

One of the more common ways in which digital inclusion is articulated has been through the "digital divide." Often seen as a binary state (someone has a computing device or they do not, they have access to affordable broadband services or they do not, etc.), the cumulative divide between social, economic, and geographic groups can seem overwhelming to bridge. At the same time, it conjures up the idea that there is a simple technological fix, that is, expanding technology itself is the bridge, for example, that a highly dynamic divide can be bridged. However, more and different technologies emerge every day, which can result not only in a larger divide and/or multiple divides, but one that will exist so long as new technologies surface and groups are left behind. Rather than viewed as a binary, thinking in terms of gradation and degree, having the capability to participate becomes the focus (Warschauer, 2003). Thus, a continuum may be a more useful way to make digital inclusion attainable (a point informed by Greenwood and Agarwal, 2016).

An example of just such an approach is the Digital Access Continuum (Figure 9.1). A group of researchers and educators working in the Cooperative Extension Service collaborated to develop this model. Organized through the National Digital Education Extension Team (NDEET) and the Extension Committee on Organization and Policy's Broadband and Digital Skills Program Action Team, it provides a useful way for thinking about advancing digital capital toward greater social inclusion. It treats each part of the continuum as necessary for advancing the quality of life. Being aware of how broadband can be used as a resource is essential but also must be coupled with those services being available (including physically available and affordable). To be able to utilize digital technologies, people need the necessary devices to make use of the connections, and they must adopt broadband. In order to advance and improve the quality of life, broadband and associated devices must be combined for actual application. This can be valuable in itself, but even more importantly, it can be used to build and exchange other forms of capital, as suggested in the SLA and CCF.

One of the somewhat unique challenges faced by groups and communities historically underserved by Internet service providers in expanding and sustaining broadband services is the need to build a critical mass of subscribers to the services to make them economically viable in the long run. Affordability is essential but so too is accepting the value proposition that utilization is valuable

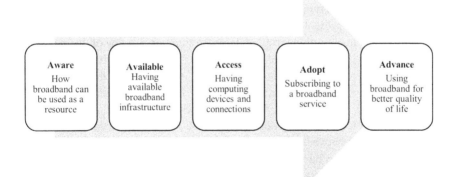

FIGURE 9.1   The digital access continuum. Image by authors.

enough for the individual, family, and business to drive adoption. This requires expanding awareness through outreach, education, training, and technical support/assistance to foster digital skills. This may be a particularly appropriate role for community empowerment intermediaries, such as community-based organizations, non-profits, colleges and universities, and the Cooperative Extension System. As suggested by Whitacre (2008, Conclusions and Implications section), "encouraging broadband demand among sectors with historically low adoption rates" may be particularly appropriate for Extension professionals.

Digital Capital Survey

To inform the development of Extension education programming, the authors were part of a team that conducted a survey to document and better understand people's digital capital and position on the Digital Access Continuum. Importantly and intentionally, the survey was primarily conducted with people already having some basic level of access to and use of computing devices and Internet services. Most data were collected through an online survey, with an additional paper response option among people pursuing digital education.

In late spring/early summer 2022, respondents were recruited via three approaches: an online panel, oversampling individuals with specific characteristics (minorities, older age, less educated, and lower income individuals); a link to the survey sent through online channels for distribution, including a national network of Extension professionals, clientele, and other stakeholders; and research team members distributed and gathered several paper copy responses. There were 968 valid responses (including 600 from the online panel, 326 via shared link, and 42 paper copy responses). Responses were received from people across 43 states and DC, including the 13 states in the southern region. For analysis, data were weighted by age to align with 2020 Census demographic characteristics.

A descriptive report of findings from the survey was published (Gallardo and Green, 2022) and presented through an online seminar for Extension professionals seeking to advance digital capital. Here we summarize a few of the highlights (Table 9.1). While most respondents had a home Internet subscription, those who did not cited cost, speed, and reliability as key barriers. Of those with a connection, cable was the most common, and only 10% had access to fiber.

There were substantial differences in these availability and accessibility characteristics between groups (Table 9.2). For example, there was a 22.8 point difference between the youngest and oldest age groups, with younger groups less likely to have a paid home subscription to Internet services. As a group, minoritized racial and ethnic groups were 14.8% less likely than non-Hispanic whites to have a home subscription, and there were differences in both education (19.2 percentage point difference between those without a high school education and those with a bachelor's degree or higher) and income groups (20.5 percentage point difference between those with an income less than $35,000 compared to those with $75,000 or more).

In terms of how survey respondents might use their digital access to potentially improve quality of life, Figure 9.2 shows digital engagement for information and services, listed in order of prevalence. As this survey was conducted in 2022 amidst the continuing challenges of COVID-19, it is not surprising that the most frequent digital engagement was for health care. Interestingly, this was followed in prevalence by local businesses and schools. Of the options provided to respondents, engagement with government, government services, and non-local businesses were the least frequently reported. In all, the findings demonstrate the potential for increased digital engagement and utilization, and they highlight

TABLE 9.1 Digital Capital Survey Respondents' Digital Availability, Access, and Adoption

Digital Access	Percent
Paid home Internet subscription	($n = 968$)
Yes	73.7
Yes, but not for the full 12 months	13.3
No	12.9
Reasons for no home Internet (agree/strongly agree)	(n, range $= 116$–120)
Not available	28.0
Too expensive	47.9
Too slow	35.4
Not reliable	39.1
Use smartphone	59.4
Type of home Internet connection	($n = 843$)
Dial-up or satellite	10.4
Cellular data	10.0
DSL	13.6
Fixed wireless	3.1
Cable	46.4
Fiber-optic	10.1
Not sure	6.4
Monthly Internet costs (Internet only, no bundles)	($n = 521$)
Less than $30.00	17.6
$30.00–$49.99	25.2
$50.00–$74.99	31.8
$75.00–$99.99	16.0
$100.00 or more	9.4

Source: 2022 Digital Capital Survey, data weighted by age groups.

TABLE 9.2 Paid Home Internet Subscriptions by Group

Paid Home Internet Subscription (%)	Yes	Yes, But Not for the Full 12 Months	No
Overall (*n* = 968)	73.7	13.3	12.9
Age 18–34 (*n* = 288)	62.2	20.5	17.4
Age 35–64 (*n* = 475)	76.2	12.0	11.8
Age 65 or older (*n* = 200)	85.0	6.0	9.0
Less than $35,000 (*n* = 627)	68.7	15.2	16.1
$35,000–$74,999 (*n* = 220)	80.9	12.3	6.8
$75,000 or more (*n*=111)	89.2	5.4	5.4
High school or less (*n* = 565)	68.3	14.7	17.0
Some college (*n* = 207)	76.3	14.5	9.2
Bachelor's or higher (*n* = 192)	87.5	7.8	4.7
White, non-Hispanic (*n* = 453)	81.7	10.2	8.2
Minorities (*n* = 504)	66.9	16.3	16.9
Metro (*n* = 756)	73.7	13.8	12.6
Non-metro (*n*=210)	73.8	11.9	14.3

Source: 2022 Digital Capital Survey, data weighted by age groups.

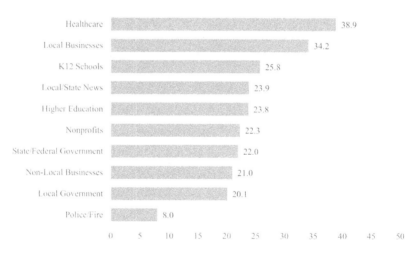

FIGURE 9.2 Share of respondents engaging digitally for information and services. Image by authors.

Source: 2022 Individual Digital Capital Survey, data weighted by age groups (*n* = 968).

opportunities for local organizations to increase their online presence. For instance, with just over one-third of respondents noting digital engagement with local businesses, there is room to expand and better connect with customers.

Given concerns with digital inclusion, an index was developed to measure the number of organizations and institutions that respondents engaged with digitally

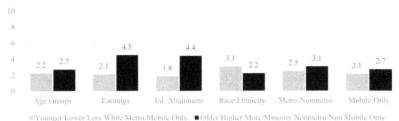

FIGURE 9.3   Average number of organizations digitally engaged with by group. Image by authors.

Source: 2022 Digital Capital Survey, data weighted by age groups (*n* = 968).

(Gallardo and Green, 2022). The scores had a possible range of 0–10, with an overall average of 2.6. Figure 9.3 compares the averages between groups. Note that for comparisons with three or more categories, only the low/high or younger/older groups were included in the graph for the sake of space.

For age, older respondents (ages 65 or older) interacted with a higher number of organizations compared to younger respondents (average of 2.7 compared to 2.2). The largest differences between groups occurred among lower and higher earning individuals (2.1 versus 4.5) and less and more educated (1.8, 4.4). Whites engaged more compared to minoritized groups (3.1, 2.2) while non-metro areas engaged more on average compared to metro areas (3.1, 2.5). Lastly, mobile-only users engaged less compared to non-mobile-only respondents (2.11, 2.70). Most of these patterns follow what we might expect for digital engagement within the broader population (education, earnings, race/ethnicity, and mobile-only), others may be surprising (age and non-metro). However, it is important to keep in mind that the survey was conducted with people who were already using the Internet/computing devices at some level.

Survey respondents were also asked about their interest in a variety of different digital skills training opportunities. While some reported not being interested in any training, more than one-quarter of respondents reported interest in the following topics: communicating with others, managing and paying bills, accessing resources, accessing entertainment, advancing career/work, and improving education.

Community Empowerment Efforts through Extension

The data reported in this chapter were used to help inform the development of the Digital Skills Building Toolkit (a training curriculum) and the Digital Volunteer Initiative (DVI). The Southern Rural Development Center (SRDC) worked with the NDEET and the Extension Committee on Organization and Policy's Broadband and Digital Skills Program Action Team members to compile, adapt, and expand existing training efforts and develop new materials to assist local investments in digital capital. With support to further develop and pilot such efforts, a team of Extension professionals associated with historically Black Land-Grant Universities (often referred to as 1890 institutions because of the year of their authorizing legislation) contributed to these efforts and have played important roles in their implementation, especially in challenging social and economic contexts in the southern US. The team was in communication with other efforts within the Extension System (such as 4-H mentioned below) and the National Digital Inclusion Alliance (so that the Digital Volunteer discussed below complemented NDIA's Digital Navigator program).

Digital Skills Building Toolkit

Twelve digital skillsets were identified as necessary for those relatively new to broadband to advance their quality of life through the resources available online. These include resources to support financial well-being, business viability, career development, education, healthcare access, and social/emotional connections. These digital skills content areas were synthesized into the Digital Skills Toolkit. The toolkit's primary focus was on knowledge and skills needed to strengthen the adopt and advance phases on the Digital Access Continuum (Figure 9.1). Creating a learner pathway through toolkit utilization, applicable to Extension program areas, allowed tailoring of relevant pathways specifically for each community and its needs.

A multi-step process of cataloging existing tools and curricula was crucial for maximizing the efficiency of resources. Through identifying materials already available, the team avoided duplication of effort, streamlining the development process, and allowed focused efforts that truly required added resources. It also allowed Extension Educators to create a sequence to provide a logical and safe entry point into the digital world. Examples included farmers learning digital applications for agriculture, workforce development exploring online job search navigation, secluded rural communities gaining access to health information and services, and senior populations being able to ease social isolation. This "pathways" approach to utilizing the Digital Skill Toolkit, allows Extension to meet people where they are, so participants progress naturally to inclusion at a pace most conducive to local needs.

By serving as community empowerment intermediaries, Extension Educators can implement the Digital Skills Toolkit (Scammahorn and Kelly, 2023) with communities by emphasizing:

* *Training Local Extension Agents:* Preparing and empowering local Extension agents with the toolkit to help them deliver relevant programming and support.

- *Local Workshops & Webinars:* Organizing online or in-person workshops to help community members understand and apply digital capital and inclusion concepts.
- *Partnerships:* Building partnerships with local organizations, government agencies, and private sector stakeholders to support and sustain digital inclusion efforts.

As part of the Bridging the Digital Divide Project, 1890 Land Grant University Extension Educators were trained as Digital Inclusion Coaches. The coaching model provides a model framework to address digital inclusion gaps, especially in minority communities. Empowering local stakeholders with guidance from Extension professionals, fosters sustainable change and helps ensure that programming initiatives are locally tailored and community driven. Extension Educators are often seen as trusted, research-based, resources in communities, so they tend to be well-positioned to facilitate the coaching process with local stakeholders.

Digital Inclusion Coaches were trained in a day-long workshop and supplemented with online video tutorials for additional review of the Digital Skills Toolkit. The online video tutorials served a dual purpose of providing a cost-effective follow-up training mechanism while building key technology usage skills. Coaches were also provided the opportunity for one-on-one consultations with an individual on the training team, as well as monthly virtual conference calls to encourage continual growth, capacity building, and being comfortable with the toolkit materials. The coaches and leadership team were also linked with a listserv that allowed for more instant communication among the group.

Upon completion of training, coaches selected at least one socioeconomically disadvantaged community with an interest in working on digital inclusion. Once identified, coaches connected with local stakeholders/partners that expressed interest in closing the digital gap, creating a place-based team. While the make-up of these teams varied, common features included county Extension staff, education, healthcare, Internet providers, elected/appointed officials, social services, business/economic development, and libraries. Each community team then determined how best to meet local needs.

Collaborating with their local teams, the coaches gained insights to assist them in identifying local needs and expanding impacts through connecting others in the community. The teams provided opportunities for achieving greater sustainability by encouraging and empowering local stakeholders, which supports long-term change. By focusing on the 'adopt' and 'advance' phases of the Digital Access Continuum, local efforts were intended to create a new norm of community broadband usage and implementation – creating an environment to support increased community quality of life. Through the coaches and their teams, community-based approaches were formed to support learning that binds the community together across multiple avenues of community life and cultural divides.

The Digital Skills Toolkit is now publicly available to anyone who wants to use it, and it has been shared with Extension professionals across the US. Train-the-trainer workshops have been offered, but are not required. People teaching from the Toolkit are asked to use the evaluation forms on a volunteer basis. From those opting to turn in evaluation data to-date in 2024, the Digital Skillsets have been delivered both virtually (7%) and in-person (93%) across three Southern states. A total of 21 separate sessions were taught at the community level, with each of the 12 skillsets listed below being taught at least once. These skill sets included: Buying and Selling Online; Computer Basics; Connecting to Government Resources; Internet Safety, Ethics, and Credibility; Introduction to E-Mail; Introduction to WORD; Online Etiquette; Saving, Recovering, Sharing Data and Files; Social Media Savvy; Zoom Basics Host; Zoom Basics Participant; and Using a Smart Phone or Tablet.

The top three courses attended were: Internet Safety, Ethics, and Credibility; Computer Basics; and Introduction to E-Mail. Females (72%), African American (53%), and people aged 60 or over (50%) were the largest population groups to attend. When participants were asked what they planned to do with their new skills, 58% wanted to connect more with family, friends, and customers, 47% planned to further their education, and 46% wanted to protect their confidential information when using the Internet.

One participant shared, "Great session! I know much more on social media. I better understand how to prepare for daily scams/false information. It was amazing to further understand social media." Another stated, "I understand the importance of creating strong passwords that you can remember." A participant from the Internet Safety skill class shared, "It was great to discuss all or most possible ways to stay safe online."

Digital Volunteer Initiative

The Digital Capital Survey also included questions about where community members turn for assistance with digital devices. Close to two-thirds said "family" and nearly half said "friends." Organizations and other formal types of support, including dedicated IT support, fell considerably behind these two groups. Thus, the partners identified the importance of training local volunteers to build the comfort level of those needing assistance, particularly individuals who are hesitant to adopt Internet use. These insights led to the development of the DVI.

Extension is rooted in engaging community volunteers to assist in the dissemination and adoption of research-based information. Well-known efforts include Master Gardeners and 4-H, just to name a couple. Volunteers are essential to the sustainability and success of Extension programming. Today's volunteers take on more in-depth roles to assist Educators in program implementation within their communities. To illustrate, in 2020 through the 4-H Tech Changemakers program, 400 teenagers taught tailored digital skills to 10,000 adults in 160 communities (National 4-H Council Annual Report, 2020). Utilizing this foundation, the partners expanded their collaboration to design the DVI, blending volunteer background knowledge and expertise, their understanding of their community and its culture, with their capacity to tackle the digital divide.

The DVI was intended to build on these efforts to develop an IT support model for adult volunteers. In addition to increasing broadband users' access to valuable online resources, this effort is intended to:

- Increase the percentage of adults volunteering in digital skills building within their communities;
- Grow the number of digital skills services in each community;
- Leverage Extension's educational programming;
- Convene community partners and strengthen collaboration;
- Enhance self-efficacy and/or comfort with digital skills among key groups in a community;
- Strengthen social and community connections; and
- Provide guidance on cybersecurity and secure online financial transactions (Leach et al., 2024).

Still in the infancy of deployment, the DVI provides a step-by-step planning process that prepares Extension to develop an individualized community empowerment strategy. Commencing with understanding the community allows the educator, through a community assessment, to identify the strengths and needs within the community. Assessment methods may include focus groups, surveys, direct observation, or interviews with key stakeholders. The Action Plan Guide worksheet provided in the guide assists in developing, organizing, and analyzing community data (Leach et al., 2024).

Creating community partnerships should be beneficial to all parties, advancing the work and outreach with like-minded goals multiplies potential impacts. Partners may include entities such as educational, governmental, non-profit, agricultural, and social/civic organizations, as well as specific people or small businesses within the community. Extension educators and the volunteer team consider the service area, intended audience, community resources, volunteer availability and safety, university policy, and protecting an individual's well-being, personal identifiable information, and privacy when creating a community plan of action. Exploring feasible options such as virtual, in-person, or a hybrid program delivery method helps reduce barriers to digital adoption. As the educator considers which delivery method may be implemented, they should consider what would need to be provided to support those efforts (Leach et al., 2024). The DVI model incorporates three program roles: program coordinator/educator, digital volunteers, and student volunteers. Continued professional development, including initial and ongoing training, is vital, as not all digital volunteers have the same depth and breadth of digital knowledge or teaching experience, coupled with the fact that broadband technologies, devices, and applications evolve rapidly.

To support all these efforts, an evaluation and impact toolkit was developed to help Extension gather and report success. The evaluation process is to emphasize the importance of effectively demonstrating the positive impact the DVI has on digital adoption and advancement. Collaborative efforts showcase the valuable contribution Extension services provide in narrowing the digital divide. Sharing successes will help Extension grow the program's impact by attracting more participants and volunteers, creating new partnerships, and raising awareness about how the DVI is impacting community vitality.

There are several DVI models available and worth exploring to determine which one is the best fit for the community. Each model has different core leadership, volunteer structure and expectations, and community involvement opportunities. However, at the core of each model is the desire to address how communities can be empowered to close the digital divide.

From Training to Empowerment

The development of livelihood pathways in the digital age requires attention not only to digital infrastructure and services but also to digital awareness, skills, and utilization. People need information and training to make use of ever-changing information and communication technologies. This can be conceptualized as digital capital. Recognizing digital access as a continuum rather than just a single binary divide, can help individuals, households, and communities design programs to address digital inclusion.

The Cooperative Extension Service has the potential to serve as a key community empowerment intermediary to help residents build and utilize their digital capital to improve the quality of life. The examples provided here – conducting the Digital Capital Survey and creating the Digital Skills Building Toolkit and DVI – illustrate the possibilities of community empowerment intermediaries to serve traditionally underserved groups. Other efforts, including Extension's 4-H Tech Change Makers and the National Digital Inclusion Alliance's Digital Navigators program, are similarly worth exploring. These and many other efforts advance models intended to ensure that traditionally underserved populations have the tools, knowledge, and resources they need to fully participate in the digital world. Communicating and engaging across diverse programmatic efforts, these types of community empowerment intermediaries can significantly advance digital capital.

Funding

This work was supported by funding through the US Department of Agriculture, National Institute of Food and Agriculture's Agriculture and Food Research Initiative (AFRI, #2022–68006-36496) and the Extension Foundation's New Technologies in Ag Extension Program (#342090). The Southern Rural Development Center also receives base support from the US Department of Agriculture, National Institute of Food and Agriculture (USDA NIFA).

Note

1 The Digital Skill Building and Digital Volunteer Initiative can be accessed here https://srdc.msstate.edu/programs/digital-skill-building-and-digital-volunteer-initiative.

References

Beaulieu, L.J., and Cordes, S. (2014). Extension community development: Building strong, vibrant communities. *Journal of Extension,* 52(5). https://tigerprints.clemson.edu/cgi/viewcontent.cgi?article=2328&context=joe

Cafer, A., Green, J.J., and Goreham, G. (Eds.). (2024). *More than Bouncing Back: Examining Community Resilience Theory and Practice.* London: Routledge.

Chambers, R., and Conway, G.R. (1992). Sustainable rural livelihoods: Practical concepts for the 21st Century. *IDS Discussion Paper* #296. https://www.ids.ac.uk/download.php?file=files/Dp296.pdf

De Haan, L.J. (2012). The livelihoods approach: A critical exploration. *Erdkunde,* 66(4), 345–357. https://doi.org/10.3112/erdkunde.2012.04.05.

De Haan, L.J., and Zoomers, A. (2003). Development geography at the crossroads of livelihood and globalization. *Tijdschrift voor economische en sociale geografie (Journal of Economic and Human Geography),* 94(3), 350–362. https://doi.org/10.1111/1467-9663.00262

Emery, M., and Flora, C. (2006). Spiraling-up: Mapping community transformation with the community capitals framework. *Community Development,* 37(1), 19–35. https://doi.org/10.1080/15575330609490152

Ezell, S. (2021). Assessing the state of digital skills in the U.S. economy. Information Technology & Innovation Foundation. https://itif.org/publications/2021/11/29/assessing-state-digital-skills-us-economy/

Flora, C.B., Flora, J.L., and Gasteyer, S.P. (2016). *Rural Communities: Legacy and Change,* fifth edition. New York: Routledge.

Gallardo, R., Beaulieu, L.B., and Geideman, C. (2021). Digital inclusion and parity: Implications for community development. *Community Development,* 52(1), 4–21. https://doi.org/10.1080/15575330.2020.1830815

Gallardo, R., Collins, A., and North, E. (2018). Community development in the digital age: Role of extension. *The Journal of Extension,* 56(4), Article 26. https://doi.org/10.34068/joe.56.04.26

Gallardo, R., and Green, J.J. (2022). Bridging the Digital Divide in Socio-economically Disadvantaged Communities in the South: Individual Digital Capital Survey Results. Mississippi State: Southern Rural Development Center. https://srdc.msstate.edu/sites/default/files/2023-04/individual-digital-capcity-survey-report.pdf

Green, J.J. (2024). Rural development in the digital age: Exploring information and communication technology through social inclusion. *Rural Sociology*, 89(2), 185–194. https://doi.org/10.1111/ruso.12542

Greenwood, B.N., and Agarwal, R. (2016). Matching platforms and HIV incidence: An empirical investigation of race, gender, and socioeconomic status. *Management Science,* 62(8), 2281–2301. https://doi.org/10.1287/mnsc.2015.2232

Gutierrez-Montes, I., Emery, M., and Fernandez-Baca, E. (2009). The sustainable livelihoods approach and the community capitals framework: The importance of systems-level approaches to community change efforts. *Community Development*, 40(2), 106–113. https://doi.org/10.1080/15575330903011785

Leach, K., Seals, L., Smyer, A. & Welborn, R. (2024). Extension Digital Volunteer Initiative Guide. National Digital Education Extension Team ECOP Broadband & Digital Access Program Action Team. https://srdc.msstate.edu/sites/default/files/2024-03/Extension-Digital-Volunteer-Program-Guide-March-2024.pdf

Muro, M., Liu, S., Whiton, J., and Kulkarni, S. (2017). Digitization and the American Workforce. Brookings. https://www.brookings.edu/wp-content/uploads/2017/11/mpp_2017nov15_digitalization_full_report.pdf

National 4-H Council. (2020). 2020 Annual Report. Cooperative Extension System, National Institute of Food and Agriculture, United States Department of Agriculture. https://4-h.org/about/annual-report/

Newman, L., and Dale, A. (2005). The role of agency in sustainable local community development. *Local Environment*, 19(5), 477–486. https://doi.org/10.1080/13549830500203121

Nyamwanza, A.M. (2012). Livelihood resilience and adaptive capacity: A critical conceptual review. *Jàmbá Journal of Disaster Risk Studies*, 4(1), #55. https://jamba.org.za/index.php/jamba/article/view/55

Ragnedda, M. (2018). Conceptualizing digital capital. *Telematics and Informatics*, 35(8), 2366–2375. https://doi.org/10.1016/j.tele.2018.10.006

Scammahorn, R., and Kelly, C. (Eds.). (2023). Digital skills building. Southern Rural Development Center, Mississippi State, MS. https://srdc.msstate.edu/programs/digital-skill-building-and-digital-volunteer-initiative

Van Dijk, T. (2011). Livelihoods, capitals and livelihood trajectories: A more sociological conceptualisation. *Progress in Human Geography,* 11(2). https://doi.org/10.1177/146499341001100202

Warschauer, M. (2003). *Technology and Social Inclusion: Rethinking the Digital Divide.* Cambridge, MA: The MIT Press.

Whitacre, B. (2008). Extension's role in bridging the broadband digital divide: Focus on supply or demand? *Journal of Extension*, 46(3), 3RIB2. https://archives.joe.org/joe/2008june/rb2.php

PART 4
Implications for the Future

This book has outlined the major issues in extending broadband in rural, low-income, and minority communities and has articulated the implications for future policy. We have profiled the role of state policy and local innovation. Challenges going forward include how to ensure continued collaboration – between federal, state, and local governments and private and non-profit actors. Compared to other countries, the US lags behind in digital inclusion. A new paradigm is needed. This section provides a vision for the future.

DOI: 10.4324/9781003619208-13

10

A NEW POLICY WINDOW TO CENTER DIGITAL INCLUSION

*Mildred E. Warner, Natassia A. Bravo,
and Duxixi (Ada) Shen*

When we began writing this book, the Biden Administration had succeeded in passing the American Rescue Plan (ARPA) and the Infrastructure Investment and Jobs Act (IIJA), which made possible the largest investment in bridging the digital divide in a century. ARPA allocated billions in pandemic-relief funds to states and localities, and supported long-term investment in public infrastructure and capital projects that enabled remote work, learning, and health monitoring. IIJA went several steps further. Billions were set aside specifically for last-mile broadband deployment, middle-mile infrastructure, tribal connectivity, and affordable connectivity. Furthermore, by including the Digital Equity Act in IIJA, federal policy recognized that the issue is not just a matter of infrastructure availability, nor is the nature of the divide only urban-rural. The digital divide disproportionately impacts low-income and minority neighborhoods as well, in part due to digital discrimination. The issue is multi-dimensional, and increasing access to infrastructure is only part of the solution. The Digital Equity Act provided support to state and local planning efforts to identify and address barriers to adoption and digital literacy.

The Biden Administration pursued a politics of repair (Xu and Warner, 2024), which centered equity. As we write this last chapter, Donald Trump has been reelected. The election demonstrated the fractured nature of U.S. society, which is reflected in the digital divide. Can individuals effectively participate in their democracy without access to fast, reliable, and affordable Internet service? President Trump capitalized on the disaffection of those left behind, but will he maintain investments and policy to promote digital inclusion? Now is an especially important time to study the impacts of prior policies as a roadmap for the future. Inadequate government investment and regulation have amplified the

DOI: 10.4324/9781003619208-14

market failure that created the digital divide. The solution requires government intervention. The market will not bridge this divide on its own.

Multi-level Framework for Community Resilience

What can we learn from past state and local efforts to close the divide? In this book, we have articulated a multi-level governance framework (Figure 1.1) to understand how government policy, technology, and the constellation of market and community actors affect how local communities address their digital divides. We have shown the challenge created by conflicts in state and federal policy and the need to support local and regional efforts to address the digital divide.

Why does the digital divide persist? We asked this in Chapter 1, and gave an overview of the market, regulatory, technology, and geographic barriers to universal access and digital equity. Broadband rollout is capital-intensive, and private providers will prioritize profitable areas – that is, high-density and wealthy populations. Providers have little incentive to build or upgrade their infrastructure in rural and high-poverty areas. The lack of competition allows Internet access in these areas to remain slow, expensive, and uneven. Market forces failed to close the divide, despite millions in government subsidies and industry-friendly legislation that could empower broadband oligopolies. When the COVID-19 pandemic accelerated the shift of essential activities to a virtual mode, the government intervened and shifted to a more comprehensive approach – prioritizing digital equity as well as universal access. Federal and state agencies are fostering competition by supporting small and non-traditional providers. They are adopting redistributive funding policies to assist communities with limited resources.

Communities rely on state support and market solutions, but they also have agency. Many foresaw that private investments would not be coming, and sought alternative paths. Some worked with their neighbors to pique the interest of providers and access broadband grants. They partnered with small, rural, and/or non-traditional providers. A growing number raised funds to build their own networks, despite state preemption. This book has shown the myriad ways in which local communities can act and how states can help along the way. We emphasize the need for multi-level governance that prioritizes public interests in critical infrastructure rollout. Broadband may not yet be a utility, but it has become fundamental to participate in society and the economy.

Digital equity requires attention to the process. Our case studies in Chapters 4–8 have shown what communities focus on, who they partner with, and how they blend federal and state subsidies with market players and local initiatives. Our emphasis on rural communities shows the critical importance of local coalitions, leaders, and resources. In Parts 1, 2, and 3 of this book, we showed that the local process is critical. Actors matter, and so does power.

Below, we provide a critical theoretical lens to understand the role of local actors. We emphasize the importance of centering public values in planning, as we focus on implications for broadband policy design.

Broadband Networks as Socio-Technical Systems

Large infrastructure networks like water, electricity, roads, and broadband can be studied as socio-technical systems – constituted by technical (material) and social elements (relationships). These systems are constantly expansive, capital-intensive, interlinked with other systems, and resistant to change (Sovacool et al., 2018). We focus on the social aspects of broadband infrastructure – the actors and relationships that constitute these large systems. Broadband planning and deployment involve a multiplicity of actors, whose functions often overlap. Public and private providers plan, design, build, manage, and own broadband networks; private and public entities fund and finance broadband projects; federal agencies and states regulate broadband; and customers and secondary systems rely on broadband access – including employers and anchor institutions, such as schools, libraries, and hospitals.

To achieve universal access, broadband networks must expand into remote, low-density, and high-poverty areas. The process is capital-intensive, and there is pressure on private network operators to ensure profitability. The inherent tension between universal access and profit-maximization drives the digital divide. Providers often resist public policy efforts to force them to upgrade or retire their networks: they fear the threat of competition and changing broadband definitions that make these older technologies obsolete. However, broadband networks must expand and upgrade to adapt to the evolving technology demands of users and to disruptive events (like the pandemic). Otherwise, they will simply become stagnant and decline (Hughes, 1983; Sovacool et al., 2018). This is already visible with copper-based Internet service, which no longer meets modern broadband speed definitions.

Broadband infrastructure is increasingly interlinked with other critical systems, especially "smart" infrastructure grids. As mentioned in Chapter 6, electric cooperatives are building their own internal fiber networks, which connect their substations and allow them to distribute electricity more efficiently. Some of these cooperatives are now leasing additional capacity to last-mile ISPs, entering the retail broadband market themselves, or partnering with municipal broadband networks. This opens up new opportunities in rural areas, where fiber optic buildout is not financially viable.

Power and Public Values

The "public" nature of infrastructure networks is related to their capacity to benefit societies by facilitating the circulation of individuals and goods. First used as a metaphor in 1652 by English economist William Petty, networks – such as travel routes, communication lines, and financial markets – function as "arteries" that allow the circulation of "blood" (resources, communications and capital) that nurture society (Van der Vleuten, 2004). Due to their social impact, infrastructure planning, construction, and management remain matters of public interest. Historically this calls for government leadership to guarantee universal access. Broadband is not technically a public good (because it can be provided by market actors), but it has become an essential service.

Conflict emerges due to the inherent tension between the public value of infrastructure, and the pursuit of profit in its deployment. Infrastructure networks are uniquely attractive to private investment. They are expansive, ubiquitous, and often non-competitive. Because they link locations with other systems, they can transform the landscape. They can change how we perform everyday activities that rely on these networks – for example, the change from in-person to virtual mode in education, work, and healthcare. Profit-driven infrastructural changes and uneven development are products of, and central to, efforts to facilitate the circulation and accumulation of capital. Infrastructure networks are developed as commodities – enabled by corporate capture, which often advocates for deregulation, halts competition, and can splinter access (Graham and Marvin, 2001), leaving millions behind.

Achieving universal broadband access and digital equity is a collective effort that calls for a renegotiation and redistribution of power between federal agencies, states, and local actors. Long before the pandemic raised concerns over digital exclusion, community broadband initiatives were already centering public values in infrastructure planning. Communities opted to invest in their own broadband networks and fought to keep Internet service affordable. They even established initiatives to address other digital inclusion concerns, like digital literacy and remote access to essential services. In this book, we have shown how community broadband networks seek to fill the gap between community connectivity needs and inadequate market coverage in the context of broadband policy that tends to prioritize market solutions and protect industry interests. State and federal policy can no longer neglect community leadership. It is essential to bridge the digital divide.

Still, community-led broadband initiatives face some challenges. They must interact with other market entities, and are bound by a series of federal and state policies. Community broadband networks must negotiate with other privately owned broadband networks to transmit data; with utilities to gain access to utility poles and use of the right-of-way, and with other municipalities to aggregate demand and share the costs of building a network. Strover, Riedl, and Dickey (2021) identify new policy approaches that can support community broadband initiatives. First, policy can facilitate access to federal and state resources by authorizing a broader range of providers to deliver broadband service – including municipal networks and public-private partnerships.

Second, public infrastructure values should be centered in broadband planning, including collective ownership to retain public control and an "open access" design that grants access to multiple providers. Finally, because broadband is essential, providers must be prepared to shoulder the burden of keeping service affordable and being transparent about their pricing.

Access to community resources is another critical aspect of broadband planning. Communities need leverage to play a more active role and retain some control in these partnerships to ensure that public objectives are met. While broadband is not a public utility or a natural monopoly, it remains an essential service. The ideal public infrastructure should be universally available and accessible to all individuals, interlinked with other essential services, and support positive externalities for the community (O'Neill, 2010).

Community-led broadband initiatives benefit from access to a high-skilled population, broadband advocates with technical knowledge, and volunteers with experience in grant-writing (Ashmore et al., 2017). Broadband advocates can identify community connectivity needs and pave the way to access federal or state broadband funding, which is so critical for communities with limited resources, who would otherwise remain unserved or underserved. The prospect of public broadband funding can help communities attract ISPs, or build their own network. Access to fiscal resources also facilitates matching a percentage of the project's costs, which increases a community's chances of receiving a federal or state grant. Finally, municipalities that run their own electric utilities have access to critical infrastructure which could facilitate their entry into the broadband field.

In Chapters 4–9, we learn how local officials, providers, and advocates leverage community resources and state and federal funding to improve Internet access in underserved areas. In a dynamic broadband environment characterized by change and uncertainty, resilient communities intentionally develop a collective capacity to respond to and influence change, and develop new trajectories for the community's future.

Taking Advantage of This Policy Window

The digital divide is not just a problem of market failure, it is a problem of policy failure, which is being addressed in the policy window opened since the COVID-19 pandemic. The federal government's largest investment to date is the $42.45 billion Broadband Equity, Access, and Deployment Program (BEAD), established in 2021. The National Telecommunications and Information Association (NTIA) is in charge of allocating these funds to states and outlining the broader eligibility requirements. Since its announcement, the rules have been scrutinized and hotly debated by government officials, industry leaders, and field experts. In some ways, it constitutes a departure from long-standing practices in public broadband funding.

For one, it raises the standards of service for eligible projects. The minimum speed eligible (100/20 Mbps) is higher than the FCC's previous broadband speed definition (25/3 Mbps). BEAD funds can be used for different types of broadband technologies, as long as the program's requirements are met – but states are still encouraged to prioritize fiber optic rollout (Ferraro, 2022). For some states, this has already been their strategy. As our analysis in Chapter 3 shows, most state funds in the 2014–2020 period went to fiber optic projects. Nonetheless, BEAD rules might not go far enough for critics of technology neutrality. Public funds will once again subsidize subpar technologies that are only acceptable as "temporary solutions," much slower than fiber, and will quickly become obsolete (Dawson, 2023). Also, areas that already have access to these older technologies would not be eligible for BEAD, as they would be considered "served."

BEAD also casts a wider net to attract a broader range of providers. The program is open to traditional and "non-traditional providers," including public-private partnerships, utilities, cooperatives, non-profit organizations, and local governments that provide broadband services. This encourages policy reform in states where local governments are prevented from owning and operating their own broadband networks. We hope the Trump Administration will continue to encourage non-traditional providers, as this helps expand the number of actors willing to provide service in rural communities. Ironically, it is the *lack* of competition that leads to inadequate broadband coverage and higher prices in underserved areas. Where markets fail to engage, the public-sector steps in. Indeed, five states have lifted their preemption of municipal broadband networks in the last couple of years (see Figure 7.1). Most communities will not opt for public ownership; in this book, we provide examples of both publicly and privately operated networks. Our cases show local governments often prefer to work in partnership with private and mission-driven ISPs – from Brownsville, Texas to Shenandoah, Virginia to York, Pennslyvania to the Blue Hill Peninsula in Maine, the Choctaw Reservation in Oklahoma, and to Alaska native communities. The aim of state policy should be to provide communities with options, and empower them to take a more active role in broadband planning. Throughout this book, we have showcased the power of local coalitions and partnerships to incentivize private providers and address infrastructure gaps.

We also shed light on potential challenges. As shown in Part 2, many underserved communities can be rendered ineligible for broadband funding due to anticompetitive behavior. If a provider is serving a portion of the community, or has received other federal or state funding to build there, then that portion is not eligible for funding. A split service area discourages competitors from applying for funds and entering the community. Incumbents can essentially "plant their flags," even if some households remain unserved. Without competition, communities are entirely reliant on incumbents' willingness to serve every household, upgrade their service, and keep prices affordable.

High match requirements can disproportionately impact rural providers and communities with limited resources, as demonstrated in Chapter 3. Some states are willing to cover a higher portion, or even the full amount of the costs if the project's location is rural and/or economically distressed. Some communities have used federal funds (e.g., ARPA) to provide a local match. BEAD allows waiver requests for locations where deployment is already too costly (NTIA, 2022). This flexibility is key to reducing barriers to participation in the most underserved areas.

The NTIA's encouragement of fiber deployment and participation by non-traditional broadband providers has raised alarm among large telecoms and the wireless industry. Indeed, the incoming Trump Administration may return to older approaches which privileged large telecom providers. However, we believe that BEAD presents an opportunity to address the shortcomings of past federal broadband funding programs. The cases in this book clearly show the importance of expanding competition and allowing local leadership. We can build on this policy learning to craft a more cooperative federal-state-local relationship going forward. That is the promise and the purpose of this book in sharing insights from state and local innovation.

A Model for Digital Inclusion

Connectivity Is Only the First Step

Most of our cases focused on building out the infrastructure for connectivity. But that is only the first challenge. Affordability and adoption are also important. While the Choctaw case in Chapter 8 was very focused on all three As – access, affordability, and adoption – more communities will need to address affordability and provide devices and training to encourage residents to use broadband technology. Chapter 9 shows how the digital divide varies by age, gender, race, and rurality; and it showcases a Cooperative Extension training program tailored to meet local needs. Internet use is the pathway to broader engagement in work, education, health, and social relations. We cannot leave a significant portion of our communities and residents on the other side of the digital divide. The digital divide is more than just an infrastructure challenge: it undermines participation in the key elements of community life – work, education, health care, civic participation, entertainment, and social relations. We need to build a more inclusive society, and digital connectivity is key.

Due to the magnitude of IIJA's investment in broadband deployment, we need to evaluate the societal and economic impact of broadband projects. To this end, the Quello Institute proposes a number of baseline metrics to monitor the contribution of broadband projects to digital literacy and broader community outcomes such as employment, income, economic growth, education, community health, and civic participation (Knittel et al., 2024).

Past research finds that broadband has important spillover effects on broader community outcomes. For example, high-speed broadband access has a positive effect on employment in rural areas (Lobo et al., 2020; Atasoy, 2013), even at the height of the COVID-19 pandemic (Isley and Low, 2022). Better broadband access also supports entrepreneurship (Deller et al., 2021). Broadband adoption has been linked to higher achievement among elementary and middle school students, especially low-income and minority students (Caldarulo et al., 2023). In Michigan, rural students with a better Internet connection were less likely to experience learning difficulties during the pandemic, and improved their digital literacy instead (Hampton et al., 2023). Finally, broadband access can have positive spillovers of public health. Telehealth expands access (Le et al., 2023), a better Internet connection enables people to choose higher-quality health care providers, and hospitals with better connectivity improve the quality of their services as well (Van Parys and Brown, 2024). The US is an aging society, and broadband is critical to connecting seniors to services and curbing social isolation, which has become a major health concern (Zhang and Warner, 2023).

Making Markets Work

Decades of market-oriented broadband policies have left deployment primarily in the hands of private providers, who have been reluctant to serve low-density rural and high-poverty urban areas where profitability would be lower. But the cases in this book show that these communities are assessing local connectivity needs and developing local broadband plans; forming public-private partnerships with local ISPs and electric utilities, and raising funds to provide local matches for state and federal grants. However, the local scope of action is largely defined by states, and sixteen states still preempt – to varying degrees – the ability of local governments to own, build, and/or operate and maintain a broadband network (Cooper, 2024). Many of these municipal broadband restrictions have been promoted by the industry (Holmes, 2014). Thus, municipal broadband projects in these states are discouraged or outright precluded from competing for state funding. The new federal BEAD policy explicitly authorizes municipal broadband, but states may still choose to restrict municipal leadership. In Chapter 7, we show how localities build partnerships with mission-driven ISPs to work within these state policy restrictions.

In rural communities without the interest or capacity to operate their own networks, local electric cooperatives have emerged as viable partners. Many already deploy fiber to upgrade their electric grids, and excess fiber could be leveraged as middle-mile infrastructure to be leased by rural, last-mile ISPs (Read and Gong, 2022). Electric cooperatives can also partner with telephone cooperatives, facilitating access to utility poles to which fiber conduit can be attached. We discuss the role of cooperatives in Chapters 5–7. Electric cooperatives are owned by the customers they serve, and emerged during the 1930s to deliver electricity and telephone service in rural areas. While not barred from entering the broadband market, public electric utilities can be limited by state regulation (Gonzalez, 2018). States should not restrict competition by limiting municipal and cooperative broadband delivery. Broadband is a utility, but it does not need to be a monopoly. There is space for creativity – in technology change, in market organization, in community leadership and in efforts to promote affordability and adoption.

What can we learn from experience in Europe and Asia, where prices are lower, and coverage is more comprehensive? In 2015, the Center for Public Integrity's analysis of five U.S. cities and five comparable French cities found that prices in the US were almost three and a half times higher for similar broadband service. There was greater competition in France than in the US, where many areas are essentially captured by monopolies or duopolies. Furthermore, in the US, in some areas, large telecommunications providers actually seem reluctant to intrude into each other's territories and compete (Holmes and Zubak-Skees, 2015). The median price for fiber Internet in the US was $75–$83 USD per month in 2024 (Supan, 2024). In the European Union (EU), the average price for 1 Gbp fixed Internet connection ranged between 21 and 72 euros in 2022. EU countries with relatively inexpensive services included France, Italy, Spain, and several countries in Eastern Europe. The EU trails behind Japan in affordability, but is ahead of the US (European Commission, 2024).

French economist Thomas Philippon notes that several European markets, including telecommunications, are more competitive, but argues the US has abandoned pro-competition policies (Philippon, 2019; Patel, 2019). For years, the EU aimed to foster competition through open-access models and subsidies for non-traditional providers. Several local and national utility companies are becoming involved in the provision of fiber-optic Internet service. Utilities can build and own fiber optic infrastructure, either leasing their fiber to multiple ISPs or providing retail broadband service themselves. Many EU countries provide subsidies to support utility-owned networks and utility-ISP partnerships. They also have open access and non-discrimination requirements to promote competition. While broadband is not treated as a natural monopoly, it is still regarded as an essential service like other utility services. Competition is encouraged, but goals and policies are established to protect public interests (Gerli et al., 2018). U.S. policy needs to encourage competition and protect public interests.

Promoting Digital Equity

To promote digital equity, we need more attention to process and the critical role played by public policy, and local coalitions. We also need to shift thinking about the ultimate goals of bridging the digital divide. Can we shift societal values to center digital equity? Traditionally, broadband deployment has been driven by profit-maximization. As new technologies emerge and older technologies become obsolete, the gap between low-profit areas (low density and low income) and high-profit areas (high density and high income) continues to widen.

As with any essential infrastructure, governmental support is critical. In Chapter 1, our multi-level governance framework emphasized the importance of state and federal policy in broadband deployment. At the local level, broadband advocates and local governments must leverage community resources and partnerships with local market players. Their efforts can be amplified or limited by state and federal policy, which must balance public goals and private market interests. If digital equity is the desired outcome, then public values must shape not only local planning, but inform state and federal policy as well. In Chapter 3, we show that state funding policies must take into consideration other barriers to digital equity besides broadband availability and population density. In Part 2 of this book we have shown how state funding has helped rural communities gain Internet access, and Part 3 showed the importance of collective action and technical assistance to ensure access, affordability, and adoption.

Figure 10.1 provides a diagram of our digital equity model. Building on the equity framework developed by Diaz and Warner (2024), we differentiate redistributional, procedural, and conceptual elements of equity. In this book, we have shown how community and policy action can lead to a shift in values regarding universal access, and how a shift in values regarding universal access (prompted by concerns raised during the COVID-19 pandemic) stimulated new policy action to address the digital divide. We need a conceptual shift in the US which recognizes that digital inclusion is fundamental to societal well-being. First, it is foundational to community resilience. Universal access enables essential activities and services to shift online and remain in operation during disruptive crises, such as the COVID-19 pandemic. Second, it allows for broader economic and social inclusion. Universal access enables individuals to participate in remote work and the digital economy. They can engage with their peers and participate in the online flows of communications, knowledge, and resources. For Indigenous communities, Internet access is also an issue of cultural sovereignty and self-determination, as shown in Chapter 8. Currently, the lack of universal broadband access is an opportunity cost for unserved and underserved individuals, which include the elderly, rural, low income, and minorities.

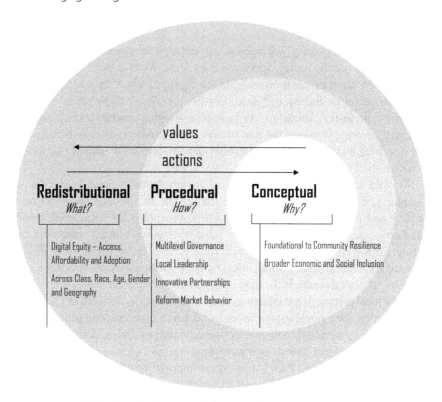

FIGURE 10.1. Digital equity framework for broadband.

Source: Image by authors, drawing from Diaz Torres and Warner 2024.

At the procedural level, the chapters in this book have demonstrated the steps that must be taken to achieve digital equity. Broadband deployment is the collective effort of actors that operate within specific geographic, socioeconomic, regulatory, political, cultural, and technology constraints. Digital equity ecosystems capture the interactions between actors and their environments which contribute to the advancement of digital equity (Rhinesmith, 2025). Our multi-level governance framework calls for policy coordination between federal and state agencies to empower communities and facilitate innovative partnerships. As Chapters 4–8 show, when traditional market solutions fail to deliver, unserved and underserved communities show their leadership by stepping up and leading the way. They form coalitions to pool their demand and resources, pique the interest of ISPs, go after state and federal funding, or share the burden of financing broadband infrastructure. They establish innovative partnerships with non-traditional market players like electric cooperatives, which facilitate access to utility poles, electric easements, and fiber optic infrastructure.

However, when local leadership is involved in broadband deployment, public and private interests will compete. State and federal funding and policy can favor public values and market reform. Public policy can promote competition by supporting the entry of small, rural, and/or non-traditional providers, for whom public funding is critical. To this end, as we show in Chapter 3, states must address barriers for grant applicants, such as prohibitive match requirements and municipal broadband restrictions. Local initiative provides communities with alternatives when traditional ISPs are unable or unwilling to serve or upgrade their infrastructure. Federal and state policy can ensure that ISPs will work with communities by raising the standards of service and providing leverage to communities. Without leverage, these communities can be "captured" by private interests, who may weaponize municipal broadband restrictions and inaccurate broadband availability maps to prevent competition and delay or deny Internet service.

Regarding redistribution, digital equity must address problems with access, affordability, and adoption across race, class, age, gender, and geography. Too many people in the US are trapped on the other side of the digital divide, and this undermines social and economic well-being for the broader community and society. Historically, broadband availability maps have narrowly focused on the number of households served by at least one ISP. These maps then determine which areas are eligible for state and federal funding. We know these maps underestimate the number of individuals actually unserved and they fail to show how these individuals are often rural, low-income, less educated, and/or minorities, people Nicole Turner Lee (2024) calls "digitally invisible." Federal policy has moved away from a narrow focus on access to recognize the importance of affordability, adoption, and digital literacy. It is not enough that high-speed Internet service is available. The digital divide is not simply a matter of coverage.

To conclude, in order to achieve digital equity, communities must also address the barriers of affordability, adoption, and digital literacy. Internet service must be universally available and affordable, and all users should have the tools, equipment, and knowledge necessary to use it. While broadband is a private good, it is essential to everyday life – therefore, it can no longer be planned as a luxury for the very few. Broadband should be approached as public infrastructure, with universal access, affordability, and service quality as primary goals.

Michael K. Powell, Former FCC Chairman and Current President and CEO of the National Cable & Telecommunications Association, noted in 2010:

> Broadband access is the great equalizer, leveling the playing field so that every willing and able person, no matter their station in life, has access to the information and tools necessary to achieve the American Dream. More and more, job listings are exclusively available online and as technology evolves nearly every occupation now requires a basic level of digital literacy with web navigation, email access and participation in social media. To that end,

Internet access and adoption opens doors to potential jobs and opportunities that would otherwise not be available to every American. Broadband eliminates so many barriers to entry for so many different people that it's actually become a barrier to entry in and of itself if you're not getting online on a regular basis. Powell, 2010 (April 5).

Eleven years later, lack of broadband access remains a barrier that disproportionately impacts vulnerable groups – as noted by Nicol Turner Lee, the director of the Center for Technology Innovation at Brookings, and author of *Digitally Invisible* (2024):

The severity of the digital divide goes beyond the usual analogy of a three-legged stool — broadband availability, affordability, and digital literacy. Policymakers must acknowledge that efforts to close the digital divide should also address poverty, geographic, and social isolation. Put simply, people not online still face systemic societal inequalities, only this time within new systems and applications like telehealth, online vaccination scheduling, or online job applications. All while still relegated to second-class, digital citizenship. Turner Lee, 2021 (December 18)

We are still waiting for universal broadband access. There are still too many people left on the other side of the digital divide. But they are no longer invisible. We see them parked outside libraries and grocery stores attempting to access the Internet for schooling, health visits, and information. They are children, seniors, people of color – but income is not the only barrier; education and access are as well, especially in rural areas and low-income urban areas. Broadband access is a market failure, but also a public policy failure as billions in public subsidies have failed to bridge the divide (Ali, 2021). We need a new approach that centers local leadership in a supportive state and federal policy environment. State and federal broadband programs must address the regulatory and funding barriers that prevent the digitally invisible from gaining access. Otherwise, the gap will remain – despite how many billions are allocated to solve the issue.

High-speed Internet access is essential to everyday life, and key to broader social and economic inclusion. Broadband access was meant to ease the barriers of time and distance, but this is only true for those who have access, can afford it, and have the skills to use it. Universal broadband access has increasingly become a matter of public interest, and in the absence of market solutions, communities are leading the way. This book illustrates the steps communities are taking to achieve digital equity, and how state and federal institutions can use regulation and funding to support their efforts. The time is now. A policy window opened with federal funding after COVID-19, and this invited localities and states to step up, build partnerships, and address the digital divide. The federal government is a crucial partner and one we hope will continue to support state and local initiative as we move forward.

References

Ali, C. (2021) *Farm Fresh Broadband: The Politics of Rural Connectivity*. Information Policy Series. Cambridge, MA: The MIT Press. https://doi.org/10.7551/mitpress/12822.001.0001

Atasoy, H. (2013) The Effects of Broadband Internet Expansion on Labor Market Outcomes. *ILR Review*, 66(2), 315–345. https://doi.org/10.1177/001979391306600202

Ashmore, F. H., Harrington, J. H. and Skerratt, S. (2017) Community-Led Broadband in Rural Digital Infrastructure Development: Implications for Resilience. *Journal of Rural Studies*, 54, 408–425. https://dx.doi.org/10.1016/j.jrurstud.2016.09.004

Caldarulo, M., Mossberger, K. and Howell, A. (2023) Community-Wide Broadband Adoption and Student Academic Achievement. *Telecommunications Policy*, 47(1), 102445. https://doi.org/10.1016/j.telpol.2022.102445

Cooper, T. (2024, September 17) Municipal Broadband Remains Roadblocked in 16 States. *BroadbandNow*. https://broadbandnow.com/report/municipal-broadband-roadblocks

Dawson, D. (2023, February 15) Let's Stop Talking About Technology Neutrality. *POTs and PANs*. https://potsandpansbyccg.com/2023/02/15/lets-stop-talking-about-technology-neutral/

Deller, S., Whitacre, B. and Conroy, T. (2021) Rural Broadband Speeds and Business Startup Rates. *American Journal of Agricultural Economics*, 104(3), 879–1174. https://doi.org/10.1111/ajae.12259

Diaz Torres, P. and Warner, M. E. (2024) A Policy Window for Equity? The American Rescue Plan and Local Government Response. *Journal of Urban Affairs*, forthcoming. https://doi.org/10.1080/07352166.2024.2365788

European Commission (2024) Mobile and Fixed Broadband Prices in Europe 2022 [Report]. https://digital-strategy.ec.europa.eu/en/library/mobile-and-fixed-broadband-prices-europe-2022

Ferraro, N. (2022) "NTIA Grants Will Fund 'a lot of non-fiber technology,' says Alan Davidson". *Light Reading*. https://www.lightreading.com/broadband/ntia-grants-will-fund-a-lot-of-non-fiber-technology-says-alan-davidson/d/d-id/780183

Gerli, P., Van der Wee, M., Verbrugge, S. and Whalley, J. (2018) The Involvement of Utilities in the Development of Broadband Infrastructure: A Comparison of EU Case Studies. *Telecommunications Policy*, 42(9), 726–743. https://doi.org/10.1016/j.telpol.2018.03.001

Gonzalez, L. (2018, January 30) Community Broadband Pushed Out of Pinetops, N.C. Institute for Local Self-Reliance. https://ilsr.org/articles/community-broadband-pushed-out-of-pinetops-n-c/

Graham, S. and Marvin, S. (2001) *Splintering Urbanism: Networked Infrastructures, Technological Mobilities and the Urban Condition*, London and New York: Routledge.

Hampton, K. N., Hales, G. E. and Bauer, J. M. (2023) Broadband and Student Performance Gaps After the COVID-19 Pandemic. James H. and Mary B. Quello Center, Michigan State University. https://doi.org/10.25335/r71b-c922

Holmes, A. (2014, August 28) How Big Telecom Smothers City-Run Broadband. The Center for Public Integrity. https://publicintegrity.org/inequality-poverty-opportunity/how-big-telecom-smothers-city-run-broadband/

Holmes, A. and Zubak-Skees, C. (2015) U.S. Internet Users Pay More and Have Fewer Choices than Europeans. The Center for Public Integrity. https://publicintegrity.org/inequality-poverty-opportunity/u-s-internet-users-pay-more-and-have-fewer-choices-than-europeans/

Hughes, T. P. (1983) *Networks of Power: Electrification in Western Society, 1880–1930*, Baltimore, MD: Johns Hopkins University Press.

Isley, C. and Low, S. A. (2022) Broadband Adoption and Availability: Impacts on Rural Employment during COVID-19. *Telecommunications Policy*, 46(7), 102310. https://doi.org/10.1016/j.telpol.2022.102310

Knittel, M., Mack, E., Nam, A. and Bauer, J. M. (2024) Assessing the Effects of the Infrastructure Investment and Jobs Act on High-Speed Internet Access, Digital Equity, And Community Development. James H. and Mary B. Quello Center, Michigan State University. https://quello.msu.edu/wp-content/uploads/2024/10/Quello-Center-IIJA-Assessment-Framework-Final-Report-20241018.pdf

Le, T. V., Galperin, H. and Traube, D. (2023) The Impact of Digital Competence on Telehealth Utilization. *Health Policy and Technology*, 12(1), 100724.

Lobo, B. J., Alam, M. R. and Whitacre, B. E. (2020) Broadband Speed and Unemployment Rates: Data and Measurement Issues. *Telecommunications Policy*, 44(1), 101829. https://doi.org/10.1016/j.telpol.2019.101829

NTIA (2022) Notice of Funding Opportunity. Broadband Equity, Access and Deployment Program. https://broadbandusa.ntia.doc.gov/sites/default/files/2022-05/BEAD%20NOFO.pdf

O'Neill, P. (2010) Infrastructure Financing and Operation in the Contemporary City. *Geographic Research*, 48(1), 3–12. https://doi.org/10.1111/j.1745-5871.2009.00606.x

Patel, N. (2019) Why Is American Internet Access so much more Expensive than the Rest of the World? *The Verge*. https://www.theverge.com/2019/11/13/20959216/thomas-philippon-economist-interview-internet-access-vergecast

Philippon, T. (2019) *The Great Reversal: How America Gave Up on Free Markets*. Cambridge, MA: Belknap Press.

Powell, M. K. (2011, May 25) Broadband Internet Keeps the American Dream Alive & Accessible. *Huffington Post*. https://www.huffpost.com/entry/broadband-internet-keeps_b_447825

Read, A. and Gong, L. (2022, March 29) States Considering Range of Options to Bring Broadband to Rural America. *The Pew Charitable Trusts*. https://www.pewtrusts.org/en/research-and-analysis/articles/2022/03/29/states-considering-range-of-options-to-bring-broadband-to-rural-america

Rhinesmith, C. (2025) *Digital Equity Ecosystems*. University of California Press.

Sovacool, B. K., Lovell, K. and Ting, M. B. (2018) Reconfiguration, Contestation, and Decline: Conceptualizing Mature Large Technical Systems. *Science, Technology & Human Values*, 43(6), 1066–1097. https://doi.org/10.1177/0162243918768074

Strover, S., Riedl, M. J. and Dickey, S. (2021) Scoping New Policy Frameworks for Local and Community Broadband Networks. *Telecommunications Policy*, 45(10), 1–13. https://doi.org/10.1016/j.telpol.2021.102171

Supan, J. (2024) How Much Does Internet Really Cost? More Than ISPs Want You to Know. *CNET*. https://www.cnet.com/home/internet/how-much-does-internet-really-cost-more-than-isps-want-you-to-know/

Turner Lee, N. (2021, December 18) Can We Better Define What We Mean by Closing the Digital Divide?. *The Hill*. https://thehill.com/opinion/technology/586396-can-we-better-define-what-we-mean-by-closing-the-digital-divide/

Turner Lee, N. (2024). *Digitally Invisible: How the Internet Is Creating the New Underclass*, Washington, DC: Brookings.

Van der Vleuten, E. (2004) Infrastructures and Societal Change. A View from the Large Technical Systems Field. *Technology Analysis & Strategic Management*, 16(3), 395–414, https://doi.org/10.1080/0953732042000251160

Van Parys, J. and Brown, Z. Y. (2024) Broadband Internet Access and Health Outcomes: Patient and Provider Responses in Medicare. *International Journal of Industrial Organization*, 95, 103072. https://doi.org/10.1016/j.ijindorg.2024.103072

Xu, Y. and Warner, M. E. (2024) Fiscal Federalism, ARPA and the Politics of Repair, *Publius: The Journal of Federalism*, 54(3), 487–510. https://doi.org/10.1093/publius/pjae019

Zhang, X. and Warner, M. E. (2023) Cross-Agency Collaboration to Address Rural Aging: the Role of County Government, *Journal of Aging and Social Policy*, 36(2), 302–324. https://doi.org/10.1080/08959420.2023.2230088

INDEX

For Product Safety Concerns and Information please contact our EU
representative GPSR@taylorandfrancis.com
Taylor & Francis Verlag GmbH, Kaufingerstraße 24, 80331 München, Germany

www.ingramcontent.com/pod-product-compliance
Lightning Source LLC
Chambersburg PA
CBHW071425050326
40689CB00010B/1980